T5-BAH-337

Trends in the Soviet Oil and Gas Industry

ROBERT W. CAMPBELL

Trends in the Soviet Oil and Gas Industry

Published for
RESOURCES FOR THE FUTURE
by The Johns Hopkins University Press
Baltimore and London

Manufactured in the United States of America
Library of Congress Catalog Card Number 76-15940
ISBN 0-8018-1870-2
Library of Congress Cataloging in Publication Data will be found
on the last printed page of this book.

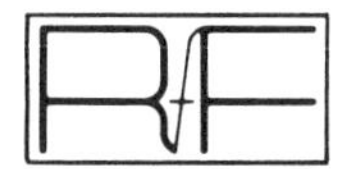

Resources for the Future is a nonprofit organization for research and education in the development, conservation, and use of natural resources and the improvement of the quality of the environment. It was established in 1952 with the cooperation of the Ford Foundation. Part of the work of Resources for the Future is carried out by its resident staff; part is supported by grants to universities and other nonprofit organizations. Unless otherwise stated, interpretations and conclusions in RFF publications are those of the authors; the organization takes responsibility for the selection of significant subjects for study, the competence of the researchers, and their freedom of inquiry. This book is a product of RFF's Energy and Materials Division, directed by Hans H. Landsberg.

The book was edited by Joan Tron.

RFF editors: Herbert C. Morton, Joan R. Tron, Ruth B. Haas, Jo Hinkel, Sally A. Skillings

FOREWORD

When Resources for the Future published Robert Campbell's *The Economics of Soviet Oil and Gas*, my predecessor Sam Schurr explained in a Foreword, dated December 1967, that ". . . the widespread interest in Soviet oil in recent years is based upon the competitive threat perceived in the expansion of exports from that country, and its emergence as an alternative to the international oil companies in meeting the needs of importing areas such as Western Europe."

Things have changed a good deal since. Far from being considered a competitive threat, the Soviet Union has come to be looked upon as a most desirable source of supplementary supplies, both of oil and gas, and not just for Western Europe but for Japan and possibly the United States. As elsewhere in the world, the question now in the United States is not how to cope with competing supplies but how long these supplies are going to last and who will have a chance of obtaining them.

Far from diminishing U.S. interest in Soviet oil and gas, these developments have, if anything, enhanced it. Thus, when Robert Campbell inquired in the summer of 1974 whether RFF would be willing to fund an "updating" of his earlier work, we eagerly took him up on his offer, stipulating only that the resulting study should be as self-contained as feasible; that is, it should be intelligible without recourse to the 1968 book.

The author has carefully observed this requirement, and the reader will find that he need not have the earlier study handy, especially as key statistical tables from *The Economics of Soviet Oil and Gas* are collected in an appendix. At the same time it should be noted that the earlier book, with its more extensive analysis of the industry and its account of the decade from the mid-1950s to the mid-1960s, still retains its usefulness. It has been supplemented, not superseded.

One of the more fascinating aspects of this monograph is the extent to which it reveals Soviet oil and gas industry problems to have much in common with those encountered in the United States. Declining well productivity, lagging finding rates, shifts to more costly deposits, refinery mix—these and other problems will strike a familiar note to observers of the U.S. scene. Oil and gas problems seem not to be respecters of differences in social and political systems. As one contemplates the change in scenery between 1968, the publication date of Campbell's first book on the subject, and 1975, one must wonder what issues may arise by the end of this decade, and whether they will call for yet another revisit!

Washington, D.C.
July 1975

Hans Landsberg
Director, Energy and
Materials Division

PREFACE

The tightening supply and rising price of energy now worrying most of the nations of the world have intensified interest in the Soviet oil and gas industry and in its present and potential role in the world market. The United States has a special interest in the matter because of the possibility of large exports of Soviet natural gas to the United States, and because Russia has shown an interest in assistance from U.S. firms, as well as from the Export-Import Bank, in overcoming some technological problems that are hampering the expansion of Soviet energy output. Intelligent assessment by businessmen and policy makers of the decisions to be made in this area requires a solid understanding of Soviet energy policies, the size and location of potential energy resources, the costs of producing them, and the technical capabilities and characteristics of the Soviet oil and gas sector. In an earlier work supported by Resources for the Future (*The Economics of Soviet Oil and Gas*, hereafter cited as *ESOG*), I dealt in detail with many of these issues. But the many data series contained in *ESOG* and the analysis and interpretation of Soviet fuel policy in it only cover the period up to the mid-sixties. Most of the analysis and the interpretations presented in that work remain valid, but it seemed that it would be useful to produce this relatively modest work as a supplement that would update the data base provided in the original book, and review developments since 1965 on those aspects of the Soviet oil and gas industry that seem most relevant in the present context.

The goal has been to fashion a review that could stand by itself, but this is a less ambitious work than the original book in the range of topics covered and in detail. Thus, both for a discussion of some aspects of Soviet oil and gas not covered here and for fuller elaboration of the background of some of the assertions made here, the reader may find it useful to consult *ESOG* along with this book. To make this a more nearly independent effort, some of the basic tables from *ESOG* covering years before 1965 are reproduced in an appendix. Insofar as possible the extensions have been made consistent with the original tables, though in some cases, the availability of new material or changes in Soviet reporting procedures has required alterations in table format.

I would like to express my appreciation to RFF, which provided funds for student assistance in preparing the new tables, to Judy McKinney and Steve Able for their help in carrying out the work, and to Tina Keller for secretarial assistance.

March 1975

Robert W. Campbell
Indiana University

CONTENTS

TABLES

FIGURES

ABBREVIATIONS FOR SOURCES FREQUENTLY CITED*

EG	*Ekonomicheskaia Gazeta (The Economic Newspaper)*
ESOG	*The Economics of Soviet Oil and Gas,* by Robert W. Campbell
GNIG	*Geologiia nefti i gaza (The Geology of Oil and Gas)*
GP	*Gazovaia promyshlennost' (The Gas Industry)*
Khim	*Khimiia i tekhnologiia topliv i masel (Chemistry and Technology of Fuels and Lubricants)*
Nar khoz	*Narodnoe khoziaistvo SSSR (National Economy of the USSR)*
Nar khoz RSFSR	*Narodnoe khoziaistvo RSFSR (Economy of the RSFSR)*
NKh	*Neftianoe khoziaistvo (The Oil Industry)*
NIG	*Neft' i gaz (Oil and Gas)*
PKh	*Planovoe khoziaistvo (Planned Economy)*
VS	*Vestnik statistiki (Herald of Statistics)*
VE	*Voprosy ekonomiki (Problems of Economics)*
ENP	*Ekonomika neftianoi promyshlennosti (Economics of the Oil Industry)*

* *The Soviet sources listed here are periodicals.*

Trends in the Soviet Oil and Gas Industry

FIGURE 1. MAJOR ECONOMIC REGIONS OF THE USSR

1

REVIEW OF SOVIET ENERGY POLICY

Before the mid-fifties, Soviet policy emphasized coal and hydroelectric power as the major primary energy sources, neglecting oil and gas. In the early fifties the USSR was a net importer of fuel, hydropower was a more important primary energy source than natural gas, and firewood contributed more toward the total energy supply than oil. Toward the end of the fifties, however, as Soviet economic planners became aware of their very large oil and gas potential, and observed the shift away from coal that had taken place in other countries, they began a sharp and accelerated development of their oil and gas resources, both to meet the growth in domestic demand for energy and to provide export earnings.

This policy has continued to the present. But there are now numerous obstacles to continuing the expansion of oil and gas outputs at the rapid rates achieved in the past, and there is some motivation to reconsider some elements of the policy. In particular, the emphasis in export policy has shifted away from oil toward gas, there is some controversy as to whether it is wise to export energy at all, there is a strong emphasis on nuclear power as an alternative primary energy source to oil and gas, and coal is once again seen as a major contributor toward meeting growing energy needs.

Major Trends Since 1965

The growth of energy production and consumption in the USSR since 1965 reveals several interesting trends, some of which are shown in table 1.

1. The rate of growth of energy output has been falling. Primary energy production grew at over 7 percent per year in the first half of the sixties but, since 1965, has grown at an annual rate of only about 4.9 percent. Year-to-year growth has been very erratic in the seventies, but overall it has been below the rate for 1965–70. Consumption and

TABLE 1. SOVIET FUEL AND ENERGY PRODUCTION, TRADE, AND APPARENT CONSUMPTION

(million tons of standard fuel)

	Production					
	Mineral fuels					
Year	Coal	Oil	Gas	Peat	Shale	Total
1965	412.5	346.4	149.8	17.0	7.4	933.1
1966	420.1	379.1	170.1	24.4	7.5	1,001.2
1967	428.6	411.9	187.4	22.4	7.5	1,057.8
1968	428.7	442.1	201.2	18.3	7.6	1,097.9
1969	439.6	496.6	215.5	16.7	8.0	1,176.4
1970	432.7	502.5	233.5	17.7	8.8	1,195.2
1971	444.2	537.3	250.6	16.7	9.5	1,258.3
1972	459.8	572.6	264.6	21.2	9.9	1,328.1
1973	468.8	613.5	282.0	20.2	10.6	1,395.1
1974	480.0	656.3	311.9	21.0	10.0	1,479.2
1975 (Plan)	491.4	699.8	342.0	n.a.	n.a.	n.a.
1975 (9th FYP)	483.5	722.8	381.9	24.6	11.5	1,624.3

Notes: n.a. = not available. The standard unit of account for fuel in Soviet planning and statistical practice is a ton of "standard fuel" defined as 7 gigacalories or 27.8 × 10^6 Btu. The average conversion coefficients in 1972 were as follows: 1 ton of coal = .702 tons of standard fuel; 1 ton of oil = 1.43 tons of standard fuel; 1,000 m^3 of gas = 1.195 tons of standard fuel; 1 ton of oil shale = .34 tons of standard fuel; 1 ton of peat = .35 tons of standard fuel.

Sources: Data, except as noted, are taken from standard statistical sources. For 1974, peat, shale, and firewood are estimated; coal, oil, and gas are from the annual plan fulfillment report in *EG*, 1975:5, hydropower is from *EG*, 1975:2.

exports have roughly paralleled the growth in production, showing the same general downward trend and year-to-year fluctuations. There is an interesting disparity in 1970, when the Russians sacrificed domestic consumption more than exports.

The decline in the rate of growth of consumption since 1965 is probably attributable mostly to a decline in the rate of growth of GNP, though this cannot be the total explanation. The sixties in general saw a lower growth rate of energy consumption than the fifties. One of the most important forces conditioning this deceleration was a shift to more efficient fuels and fuel saving processes; for example, substitution of higher-grade fuel for firewood in the household sector, a shift from steam to diesel and electric traction on the railroads, and a reduction in the fuel rate in electric power generation through technological improvements and the replacement of low-grade solid fuel with oil and gas. But it is difficult to imagine that these influences could have been more important in the second half of the sixties than in the first half. Indeed, some of these influences (for example, the shift in railroad traction)

Production				
Firewood[a]	Hydro-power[b]	Total energy	Net exports	Apparent consumption
33.5	33.8	1,000.4	101.5	898.9
31.9	37.2	1,070.3	114.6	955.7
30.6	34.9	1,123.3	123.4	999.9
28.7	40.0	1,166.6	134.8	1,031.8
28.0	43.4	1,247.8	144.3	1,103.5
26.6	45.6	1,267.4	153.6	1,113.8
26.6	45.3	1,330.2	150.8	1,179.4
25.7	43.5	1,397.3	145.9	1,251.4
25.1	43.5	1,463.7	166.6	1,297.1
25.0	50.4	1,554.6	165.0	1,389.6
n.a.	n.a.	n.a.	n.a.	n.a.
14	56.2	1,694.5[c]	n.a.	n.a.

The targets originally set for 1975 in the Ninth Five Year Plan (FYP) are from Baibakov, 1972; the adjusted targets set in 1974 for 1975 are from *EG,* 1974:52.

[a] Does not include wood gathered by the population for its own use, estimated in Savenko and Shteingauz, 1971, p. 146, as about 40 million tons in both 1965 and 1970.

[b] Converted at the fuel rate for thermal central stations of the corresponding year.

[c] In addition, nuclear plants are to supply the equivalent of 8.5 million tons of standard fuel.

were more or less exhausted by the end of the sixties. It might have been expected that the growth rate of energy consumption would pick up again once many of these changes had been made, but it seems that it has not done so.

2. Exports absorb a large share of Soviet energy production—slightly over 10 percent. But the net trade position differs among different energy sources. Thus far there has been a small net *import* of gas, a small net *export* of electric power, and a significant net *export* of coal (a little under 3 percent of output). Most of the energy export is accounted for by oil: in the 1965–74 period nearly 30 percent of Soviet oil output was used for export purposes rather than to satisfy domestic consumption (see table 25).

3. The trend toward liquid and gaseous hydrocarbon fuels at the expense of solid fuels has continued since 1965. The check to growth of coal output experienced in the first years of the sixties was shortlived, but oil and gas have consistently grown faster than other sources and have, therefore, increased their share in total energy production. The

share of oil in total primary energy production rose from 34.6 percent in 1965 to 42.2 percent in 1974, that of gas from 15.0 percent to 20.1 percent. The Ninth Five Year Plan (FYP) envisaged further increases—to 42.4 percent for oil and 22.4 percent for gas. But gas growth has fallen significantly behind plan, and its share was probably about 21 percent in 1975.

Since the various energy sources continue to play about the same relative roles in exports as they have in the past, the shift to oil and gas in production is paralleled by corresponding changes in consumption. When oil and gas play so large a role in the fuel balance, both are necessarily being used to a large extent for boiler and furnace fuel, especially in electricity generation. Crude oil is only lightly refined, with a large share of refinery output in the form of residual fuel oil, which the Russians call *mazut*. Electric power stations in 1970 consumed about 74 million tons, standard, or 53 million tons, natural, of "liquid fuel" (Pavlenko and Nekrasov, 1972, pp. 170–171). Some of this (perhaps 2 million tons) was diesel fuel. The total represents a large growth from 8.4 million tons, natural, in 1960. *Mazut* burned in other kinds of boilers in 1965 is reported as 10.7 million tons, standard and 7.6 million tons, natural (Savenko and Shteingauz, 1971, p. 162). Evidence in the statistical yearbooks and Pavlenko and Nekrasov, 1972, suggests that it rose to about 10 million tons by 1970. The total amount of *mazut* burned in boilers in electric power stations and elsewhere in 1970 was thus 60–65 million tons.

By now (1975), this figure is surely very much larger. The new electric generating capacity being added in the European part of the RSFSR during the Ninth FYP, constituting over half of all the new thermal capacity for the country as a whole, was to be fueled primarily with *mazut*, with supplemental use of gas in the summer (Pavlenko and Nekrasov, 1972, pp. 174, 219). As a rough guide to the amounts of *mazut* involved, if half of the increment in power output from thermal stations between 1970 and 1975 were generated with *mazut*, this would be an additional 20 million tons or so beyond the 60–65 million tons calculated above for 1970.

From the beginning of the introduction of gas into the fuel balance, a large share has gone to electric power generation. In 1970 the Ministry of Electric Power burned 24.4 billion cubic meters of gas in its stations, and an additional 12.3 billion cubic meters were used in stations controlled by industrial enterprises. Together, these uses totaled 30.6 percent of all gas consumption. Another 20 percent was used to raise steam in nonelectric-power boilers (Umanskii and Umanskii, 1974, pp. 106–107).

4. Oil and gas have continued to provide most of the increment in total energy output. Under the targets set in the Ninth Five Year Plan, they would have accounted for 608 million tons out of a total increase of 703 million tons of standard fuel, or 86 percent, though in actual growth their share has been slightly less than this. Their role in the growth of energy consumption is approximately the same. Finally, the Soviet foreign trade handbooks show that oil and gas account for 90 percent of total energy export.

5. The one significant new development in the Soviet energy balance since 1965 is the entry of nuclear energy. It was still a negligible element in total primary energy supply in 1970, but output is now growing fairly rapidly. Nuclear energy accounted for about 16 billion kilowatt-hours in 1974, with a target of 25 billion kilowatt-hours by 1975. It is, thus, at least on the scoreboard, with one-half of 1 percent of all primary energy production.

THE ECONOMIC RATIONALE OF SOVIET ENERGY POLICY

Soviet energy policy is based primarily on the fact that oil and gas are the cheapest sources of energy both to produce and to use. But it is also strongly influenced by transport considerations. Soviet energy resources are in general poorly located with respect to areas of consumption. About 80 percent of the consumption of energy in the USSR is in the European areas, but most of the energy reserves (almost 90 percent) are located in the Asiatic areas of the USSR (Oleinik [ed.], 1972, pp. 131–132). As can be seen from table 2, a number of important economic regions (especially the Center, Ural, and Northwest regions)[1] are heavily dependent on importation of energy in various forms from other regions. Until recently these import needs could largely be met from other regions in the European part of the USSR, especially the Transcaucasus, North Caucasus, and Volga. Though I have been unable to find a comprehensive statement of the regional distribution of fuel production and consumption later than for 1965, all the evidence points to an exacerbation of the locational imbalance shown in table 2. Between 1965 and 1970 the imbalance between supply and requirements in the European part of the USSR grew, and spread to more regions as some of the traditional surplus areas were unable to increase output at a rate equal to the increase in their consumption. The Transcaucasus surplus fell, the Ukrainian SSR kept its surplus about level and the North Caucasus had a small increase. The Volga area was the one area

[1] For location of economic regions, see the map facing page 1.

TABLE 2. REGIONAL DISTRIBUTION OF FUEL PRODUCTION AND CONSUMPTION, 1965 (million tons of standard fuel)

Region	Production	Consumption	Surplus (+) or deficit (−)
Northwest, and European North	21.46	50.76	− 29.30
Center[a]	20.52	87.37	− 66.85
Volgo–Viatka[a]	1.87	22.47	− 20.60
Central Chernozem[a]	0.0	24.13	− 24.13
Volga	251.94	69.90	+182.04
North Caucasus	101.71	35.78	+ 65.93
Ural	55.05	105.68	− 50.63
Western Siberia	86.77	58.25	+ 28.52
Eastern Siberia	30.79	44.10	− 13.31
Far East	19.60	28.29	− 8.69
Ukrainian SSR	218.34	172.24	+ 46.10
West	6.53	22.47	− 15.94
Transcaucasus	38.26	24.96	+ 13.3
Central Asia	41.06	36.61	+ 4.45
Kazakh SSR	35.46	24.96	+ 10.50
Belorussian SSR	3.73	19.14	− 15.41
Moldavian SSR	0.0	4.99	− 4.99
Total	933.09	832.10	+100.99[b]

Source: Probst, 1968, pp. 45, 68.

[a] These are subdivisions of the Central region shown on the map. The Central Chernozem region is in the south, Volgo–Viatka in the northeast, and the Center is in the west.

[b] This figure represents net exports of fuel.

in the European USSR that was able to attain a significant increase in its surplus. Since the Russians seem to have been unable to change the geographical pattern of demand appreciably by shifting industry to energy surplus areas, it has become increasingly necessary to transport energy from Central Asia and Siberia to meet the energy deficit of the European USSR as a whole. By 1970 the fuel deficit of the European USSR was 140 million tons of standard fuel. Since 1970, the general European deficit has increased still further. It was forecast (*VE*, 1971: 6, p. 55) that by 1975 Western Siberian and Kazakh coal, oil, and gas, and Turkmen and Uzbek gas would be contributing as much as 350 million tons of standard fuel to European deficit areas, and scattered evidence suggests that the figures will in fact be even larger than that. This is a significant proportion of a total primary energy output that was expected to be about 1,700 million tons of standard fuel in 1975.

The depletion of European resources applies to gas and oil as well as energy resources in general. The oil regions most convenient for supplying the European market areas are being seriously depleted, and there is a serious regional imbalance in the proved and probable reserves from which replacement output can come. As explained later, progress in keeping oil reserves growing along with output has been disappointing, and, at the present time, explored reserves that can support replacement production are primarily in Western Siberia. The associated transport cost seriously raises the cost of Siberian oil as a fuel in the European USSR. There are possibilities other than Siberia for replacing the big fields in the Volga–Ural regions which have, until now, been the source of oil growth (offshore deposits, deposits at great depths, more secondary recovery, use of minor fields in the present producing regions, and others), but these sources of output expansion are costly to find and use. As a result, the Russians may be forced to re-evaluate their decision to increase oil output as the cheapest way to meet the need for more energy.

The same general line of argument holds for gas as well, except that there is no reserves bottleneck. Adequate reserves to support a severalfold increase in output have already been located and explored. The problem is the cost of transporting the gas from Siberia, where the reserves are, to areas of consumption. The uncertainties here concern how rapidly these reserves can be developed, and how rapidly gas transport technology can be improved to bring them to market.

Even with these obstacles, however, the problems of transporting oil and gas from the Asiatic regions have seemed less serious than those for coal. Recently, however, the question of the relative advantage of transporting coal (possibly by long-distance transmission of electric power) rather than oil and gas seems to be once again under consideration.

Soviet fuel-policy planners have placed heavy emphasis on oil and gas production not just because this is the cheapest way of meeting the energy needs of the growing economy, but because of the attractiveness of these fuels as exports. They are standard commodities, relatively easy for the Russians to handle without elaborate marketing or service efforts. Current high prices for oil and gas on the world market have naturally enhanced their export value. Moreover, as the leader of the socialist bloc, the Russians are under special pressure to supply oil and gas to their energy-poor partners in Eastern Europe. Those countries desperately need energy imports to permit them to keep growing, which is an important desideratum for the Russians. The Russians also see the advantage to themselves of controlling so vital a commodity as a way to enhance their influence over the Eastern European countries.

For all these reasons oil and gas play a special role in USSR energy policy. In trying to predict how this policy may evolve in the years ahead, and how oil and gas will figure in it, it is clearly necessary to investigate more fully the issues of production and transport cost of oil and gas (and their potential replacements), the status of and trends in the technology that underlies these costs, and the growth of internal demand. We will return to these considerations at various points in the following chapters.

2

EXPLORATION FOR OIL AND GAS

A number of indicators related to the Soviet exploratory effort are shown in table 3. Expenditures on exploratory work for oil and gas have continued to rise in the years since 1965. Total expenditures on exploration were planned to rise in 1971–75 to 1.5 times what had been spent in the previous five-year period (Luzin, 1974, p. 26). The increase in expenditures, however, reflects cost increases more than growth in the volume of exploratory work.

A brief summary of the phases of exploration as the Russians conceive and conduct it will make more understandable the interpretation of these indicators. General geological surveys are conducted to uncover promising prospects for further exploration. Promising structures found at this stage are then prepared for deep drilling by more intensive investigation based on core drilling and geophysical methods, especially seismic mapping. The next step is to drill "prospecting" (*poiskovye*) wells to find whether the supposed structure does in fact contain oil; for those that do, the final stage in exploration is additional deep drilling to accumulate the information needed to evaluate reserves and to design a production scheme for the field. Drilling of this type is included along with "prospecting" drilling in Soviet totals for exploratory drilling.

Activity Levels

If exploratory activity is measured by the number of structures discovered and prepared for deep drilling, the volume of exploration has stagnated since 1965. The cost of preparing structures has increased as attention shifts to deeper structures, and as more work (more meters of structural drilling or more miles of seismic profile, or both) is done per structure. Total meterage of prospecting drilling has remained virtually unchanged; if the volume of drilling is corrected for changes in average depth, it appears that the number of prospecting wells declined in 1972 to less than 90 percent of the number drilled in 1965.

There is no evidence of significant change in the share of the structures prepared that turn out to be productive, and the success rate

TABLE 3. EXPLORATORY WORK IN THE SOVIET OIL AND GAS INDUSTRY

Item	Unit	1965
All exploratory work	million rubles	1,142
All exploratory drilling	million rubles	829
Geological and geophysical	million rubles	313
Geophysical	million rubles	129
Seismic	% of geophysical	41
Structures discovered	number	n.a.
Structures prepared	number	333
All exploratory drilling	thousand meters	5,565
Prospecting drilling	% of exploratory	47.5
Exploratory drilling for gas	thousand meters	1,939
Share of successful wells	% of total	
All exploratory wells		41.2
Prospecting wells only (new fields)		17.8

n.a. = not available.

Sources: Exploratory work: Figures in the table are synthetic estimates rather than a straight reporting or reconstruction of Soviet data, which was the procedure in the comparable table in *ESOG*. Hence the data for 1965 in this table differ from those for the same year in *ESOG*. Expenditures on exploratory drilling for all years are calculated as exploratory meterage times the cost per meter (from table 5). The totals and the breakdown for geological and geophysical work are based on absolute data and percentage breakdowns for 1965 and 1970, in Mel'nikov (ed.), 1968, p. 281; VNIGRI, 1967, p. 22; and Luzin, 1974, p. 14. Data for other years are interpolated and extrapolated on the basis of some data for totals for the period, indexes of growth of the two kinds of work, and relative importance of the two kinds of work, in *ENP*, 1970:1, p. 14, and 1972:11, pp. 13–17. The results are reasonably consistent with other data given in VNIGRI, 1973-a, p. 61, and in Feigin, 1974.

All exploratory work is figured as the sum of expenditures on exploratory drilling, and on geological and geophysical work.

Number of structures discovered and prepared: ENP, 1972:11 and 1973:4.

in exploratory wells has also stayed essentially stable. There may have been some decline in the success rate for prospecting wells in new fields. There was a downward trend in the first few years after 1965, and the fact that information on this indicator has not been provided for in recent years suggests a further deterioration.

Measured at the output end, exploration generates new reserves, and the amount of exploration accomplished in this sense is strongly dependent not only on the intermediate success indicators, but also on the average size of new finds. Discovery of a very large field like Samotlor, which is reported to contain reserves of 2 billion tons of oil, can make a very good year. There is no good evidence to evaluate what has happened to the average size of oil finds in the years since 1965, except that it is known that in the last several years there have been no really giant discoveries, and there is much handwringing over this. (The gas

1966	1967	1968	1969	1970	1971	1972
1,224	1,273	1,286	1,479	1,580	1,686	1,696
904	946	951	1,137	1,230	1,328	1,330
320	327	335	342	350	358	366
142	157	173	190	210	231	254
n.a.	n.a.	n.a.	n.a.	80	n.a.	n.a.
387	442	384	397	425	n.a.	n.a.
389	398	373	418	393	n.a.	n.a.
5,648	5,802	5,111	4,924	5,146	5,250	5,138
48.2	47.2	50.3	52.7	53.2	53.7	54.2
2,265	2,344	2,097	1,831	1,812	1,812	1,748
51.1	54.0	53.1	54.0	46.9	41.7	n.a.
15.8	15.2	13.5	15.2	12.2	n.a.	n.a.

All exploratory drilling: The total in meters is taken from table 5. The share of prospecting drilling in exploratory drilling is from Feigin, 1974, p. 149. The definition of this concept varies somewhat among different Soviet studies. All exploratory (*razvedochnoe*) drilling is subdivided into (1) base-point (*opornye*) wells, (2) parametric wells, (3) wells drilled into new structures, (4) wells drilled into known fields in search of new deposits, and (5) outlining wells. Thus the share of prospecting wells might be reported as the sum of categories (3) and (4) in relation to all exploratory wells, as the sum of categories (1) through (4) in relation to all exploratory wells or as the sum of categories (3) and (4) in relation to the total for categories (3) through (5). The figures reported here seem to be the share of categories (3) and (4) in the total. Total meterage of exploratory drilling for gas is from Feigin, 1974, p. 149.

Share of successful wells, all exploratory wells: 1965: N. F. Mel'nikov (ed.), 1968, p. 283; 1965–70: VNIGRI, 1974, p. 59; 1971: Luzin, 1974, p. 110. There are some inconsistencies in the various sources.

Share of successful wells, prospecting wells only: VNIGRI, 1971, p. 238, gives data through 1968, while 1969 and 1970 are given in VNIGRI, 1974, p. 35.

situation has been much more favorable, but discussion of gas discoveries and reserves will be postponed to a later section.)

Growth of Oil Reserves

As explained in *ESOG*, the Russians do not generally release absolute data on oil reserves, but the growth of several categories of oil reserves has been reported, as shown in table 4. For a detailed discussion of the Soviet reserve classification and how it fits into the exploration process, the reader is referred to *ESOG*,[1] but the concepts used in table 4 can

[1] The period since the publication of *ESOG* has produced some very enlightening discussions by Soviet authorities on the meaning of the reserve classifications, and of how the process of estimating and certifying reserves works. The issues are too technical to review here, but the essential point is that planners and officials in

TABLE 4. INDEX OF GROWTH OF SOVIET OIL RESERVES, 1961–72

January 1	A + B (1946 = 100)	A + B + C_1 (1946 = 100)	Ratio of C_1 to A + B
1961	657	225	0.49
1962	745	254	0.44
1963	770	265	0.46
1964	822	290	0.49
1965	863	306	0.50
1966	911	333	0.55
1967	950	348	0.57
1968	991	380	0.61
1969	1,027	407	0.67
1970	1,037	427	0.73
1971	1,087	457	0.77
1972	1,072	469	0.84

Sources: Feigin, 1974, pp. 26, 28; an article by the same author in *ENP,* 1968:3, p. 18; and Feitel'man, 1969. The sources are generally mutually consistent, and information therein covering the ratio of category C_1 to A+B is also consistent with these indexes.

Note that these data differ somewhat from those in table 6 in *ESOG,* which showed A+B resources on Jan. 1, 1961, as 7.65 times those of January 1, 1946, rather than 6.57 times, as here. Inconsistencies in various claims are easily understandable in view of several changes in the definitions of the various reserve categories, in the reviews of recovery coefficients, and varying determinations as to division into *balansovye* (those considered commercially exploitable) and *zabalansovye* (those considered not commercially exploitable). The series in the present table should be considered preferable to those in *ESOG.*

be briefly explained as follows. The A+B classification refers to oil that is well studied, essentially the contents of reservoirs that have been fully outlined by exploratory wells. Reserves of the C_1 category are the contents of oil bearing structures estimated on the basis of considerable knowledge of reservoir characteristics and the dimensions of the structure, if the presence of oil has actually been demonstrated by even a single successful prospecting well. The reserve classification scheme was changed somewhat in 1970, in the direction of easing slightly the requirements for inclusion of reserves in the C_1 category.[2]

The important conclusion to be extracted from table 4 is that both for the A+B and the C_1 categories, reserves grew less rapidly between 1965 and 1972—somewhat less than the 1.6-fold growth of crude oil

exploration now seem to have a much clearer idea than a few years ago of what goals the reserve classification scheme should serve, both as a basis for evaluating the activity of exploratory organizations and for coordinating the planning of exploration with the growth of oil output.

[2] This is shown by a comparison of the two classifications and is confirmed in a Soviet source: "The 1970 classification differs from the 1959 classification mainly in reducing somewhat the requirements as to the degree of exploredness of fields" (VNIGRI, 1973, p. 11).

output during the same period (see table 9). Moreover, A+B reserves have grown less rapidly than C_1, so that the share of C_1 reserves in the total appears to have risen significantly since the middle sixties. This may be due to the fact that C_1 reserves tend to be overestimated. One Soviet author says that in recent years detailed exploration of newly discovered fields has shown that the original C_1 estimates were too high, and that 40 percent of such reserves had to be written off (*ENP*, 1968: 4, p. 7).

There has been an appreciable shift in the location of oil reserves in all categories, with the share of Western Siberia having increased very greatly. Unfortunately, there do not seem to be any authoritative statistics on the current regional distribution of oil reserves, but it is strongly asserted that Western Siberia is now the dominant area.

A falling ratio of reserves to output, aggravated by the failure to find significant new reserves in areas with established production, is a bad omen for future output growth. The reasons for the tight reserve situation probably involve some miscalculations in planning. Exploration has been planned during this period on the assumption of increases in productivity of crews and equipment that have not been realized. The input allocations fixed in the plans, therefore, have meant that far fewer exploratory wells were drilled than expected and that plans for new discoveries were not fulfilled. Nearly every year officials of both the Ministry of the Oil Industry and the Ministry of Geology have reported serious underfulfillment of the goal for both A+B and C_1 oil reserve increases. At the present time oil industry spokesmen are issuing strong warnings of the need to expand reserves, and V. D. Shashin, Minister of the Oil Industry, is critical of the Ministry of Geology for failing to find the big new fields needed to keep output growing. It is also interesting to find the Ministry of the Oil Industry itself searching intensively for new fields, as indicated by a rising share of prospecting drilling in its total exploratory meterage (reported by Shashin in *NKh*, 1974:3, p. 4).

Because of the very large finds in Western Siberia during the 1966–70 period, made with relatively small expenditures of either money or drilling meterage, the "effectiveness" of exploration was very high. The absence of equally good luck since 1970 has led to a considerable deterioration of the reserves situation.

3

DRILLING

Drilling is one of the most important subsectors of the oil and gas industry. It accounts for a very large share of all investment expenditures (about 50 percent in the case of oil). Productivity trends in drilling are very important in influencing how rapidly meterage can grow, which in turn determines how fast output can expand. Some principal indicators of drilling activity are shown in tables 5, 6, and 7, which permit us to analyze activity levels, productivity, and technical progress.

Total meters drilled for oil and gas, for both exploration and development, stayed essentially unchanged in the second half of the sixties, after having almost doubled in the 1955–65 decade. It was intended that meterage would once again increase considerably during the Ninth Five Year Plan—to a total 30 percent greater than in the Eighth Five Year Plan (1966–70). The growth in meters drilled during 1971–73 was much more modest than planned, however, and it seems certain that the goal of 72.5 million meters for 1971–75 was underfilled by as much as 10 percent.

Underfulfillment of the drilling plan flows from failure to achieve intended gains in rig productivity. In exploratory drilling, the number of meters drilled per rig per month has fluctuated somewhat since 1965, as shown in table 5, but at 340 meters per rig per month in 1974 (*NKh*, 1975:8, p. 3) was well below the 377 meters per rig per month in 1965. In development drilling, this indicator has improved slowly from 1,082 meters per rig per month in 1965 to about 1,450 in 1974. The Ninth Five Year Plan envisaged quite large gains in rig productivity (70 percent for exploratory, later scaled down to 30 percent; and 50 percent for development drilling, later scaled down to 43 percent). But the performance through 1974 (no improvement in exploratory, and a 25 percent improvement in development drilling) makes clear that even the lower targets are far out of reach.

Poor rig performance and adverse conditions are reflected in rises in drilling costs, from 149 rubles to 239 rubles per meter in exploratory drilling and from 65 rubles to 85 rubles per meter in development drilling between 1965 and 1970. Average cost per meter for all drilling was supposed to fall by 20 percent in the Ninth Five Year Plan, but experience in 1971 and 1972 suggests that it is more likely to have increased by at least that much during the period.

Table 5 suggests that the Russians responded to the drilling bottleneck by focusing on development drilling at the expense of exploratory drilling. Total meterage of exploratory drilling decreased in absolute terms since 1965, and its share in total drilling fell significantly—from 52 percent in 1965 to 40 percent in 1972.

The adverse trend in rig productivity is the result of an essentially stagnant technology in the face of a more difficult drilling job. Well depth has increased and drilling has shifted to areas with harsh climatic conditions and lack of infrastructural facilities. Average depth of wells increased in exploratory drilling from 2,195 meters in 1965 to 2,622 meters in 1972, and in development drilling from 1,647 meters in 1965 to 1,781 meters in 1971.

Although much has been written about the inhospitable conditions in Western Siberia, it appears that this is only a minor factor in the overall deterioration of rig productivity and the rise in costs of drilling. The Western Siberian fields lie at relatively shallow depths, and, despite the fact that they have provided most of the gains in reserves and in output, Western Siberian fields account for a relatively small share of total drilling: less than 10 percent in the 1966–70 period (VNIGRI, 1974, p. 55). In the older areas the prospective additions to reserves and output are almost always at appreciably greater depths than current production. This is especially true in the Transcaucasus, North Caucasus, and Ukraine. As a result, it is the deterioration in the drilling environment in these traditional areas, which remain dominant in the total drilling footage, that has kept drilling productivity from rising.

Depth of Wells

There is very little comprehensive information concerning the depth distribution of wells for recent years. There are many scattered statements about the increasing importance of the deeper classes, though many of these are inconsistent and difficult to interpret, partly because the definitions of "deep" and "super deep" have been changed from time to time. But the essence of the problem is that there are now numerous situations in which wells must be drilled below 3,500 meters (11,500 feet). For the USSR as a whole, meterage in wells in the 3,500+ class rose from about 1.0 million meters in 1965 to about 1.5 million meters in 1969. This meterage was all drilled originally in exploratory well (*ENP,* 1973:3, p. 16 and Umanskii and Umanskii,

TABLE 5. SELECTED INDICATORS OF SOVIET DRILLING OPERATIONS

Item	1965	1966	1967
Meters drilled (*thousands*)	10,716	11,251	11,707
Exploratory	5,565	5,648	5,802
Development	5,151	5,603	5,905
Turbodrilling (*percent of total*)	80	n.a.	76
Drilling with electric drills (*thousand meters*)	215.1	n.a.	n.a.
Drilling with bits 9″ and smaller (*thousand meters*)	5,324	5,630	5,779
Drilling with diamond bits (*thousand meters*)	47.4	52.9	73.8
Commercial speed (m per rig per mo.)			
Exploratory	377	367	380
Development	1,090	1,137	1,107
Mechanical speed (m per hr)			
Exploratory	4.2	n.a.	n.a.
Development	n.a.	n.a.	n.a.
Holes drilled per bit (*meters*)			
Exploratory	17	16.85	17.78
Development	28.3	n.a.	n.a.
Cost of wells (*rubles per meter*)			
Exploratory	148.7	159.6	162.8
Development	65.54	n.a.	60.97
Number of wells completed			
Exploratory	2,165	2,163	2,240
Development	2,738	n.a.	n.a.
Number of new wells put into production			
Oil	2,767	3,079	3,083
Gas	n.a.	—	—
Average depth of wells (*meters*)			
Exploratory	2,195	2,423	2,447
Development	1,647	1,653	1,682
Number of rigs operating			
Exploratory	1,230	1,282	1,272
Development	394	411	445

n.a. = not available.

Sources: Meters drilled: Nar khoz, various years.

Share of turbodrilling: 1965: Mel'nikov, 1968, p. 340; 1967: *NKh,* 1969:7, p. 6; 1968: Syromiatnikov, 1970, p. 109 gives the share of rotary as 18 percent; 1970, 1972: Umanskii and Umanskii, 1974, p. 53.

Electric drilling: 1965: Mel'nikov, 1968, p. 340; 1968: Egorov, 1970, p. 40; 1969, 1970, 1972: Umanskii and Umanskii, 1974, pp. 28, 83.

Bits 9″ and smaller: Nar khoz, various years.

Diamond bits: 1965–70: Pobedonostseva, *et al.,* 1972, p. 26; 1972 Umanskii and Umanskii, 1974, pp. 81, 367.

Commercial speed, exploratory: Nar khoz, various years, except 1969–72; *NKh,* 1974:4, p. 15; 1973: *Burenie,* 1974:1, p. 6.

Commercial speed, development: Nar khoz, except 1969: *ENP,* 1970:9, p. 5; 1970: *ENP,* 1971:4, p. 5 and *ENP,* 1973:8, p. 20; 1971–73: *Burenie,* 1974:1, p. 6, *ENP,* 1973:8, p. 20.

Mechanical speed, exploratory: 1965, 1970: *ENP,* 1971:12, p. 17; 1968: *ENP,* 1970:4, p. 16.

Mechanical speed, development: 1968: *ENP,* 1970:4, p. 16.

Hole per bit, exploratory: 1965: *ENP,* 1970:9, p. 4 (given as 16.5 in *ENP,* 1971:12, p. 17); 1966–72: *NKh,* 1974:4, p. 15; 1968 given as 17.4 in *ENP,* 1970:4, p. 16.

1968	1969	1970	1971	1972	1973
11,070	11,061	11,890	12,128	12,720	13,628[a]
5,111	4,924	5,146	5,250	5,138	5,223
5,959	6,137	6,744	6,878	7,582	8,405
79	n.a.	74.5	n.a.	74.4	n.a.
300	263	273	n.a.	267	n.a.
5,466	5,491	6,019	6,091	6,739[b]	6,361
110.7	124.8	91.9	n.a.	281	n.a.
349	327	337	339	329	338
1,127	1,156	1,154	1,216	1,280	1,378
3.22	n.a.	3.4	n.a.	n.a.	n.a.
9.20	n.a.	n.a.	n..a	n.a.	n.a.
17.10	17.96	18.3	19.4	20.3	n.a.
33.2	34.1	n.a.	36.7	39.6	n.a.
185.5	230.7	238.8	253.2	259.35	n.a.
n.a.	n.a.	84.8	86.1	93.6	n.a.
2,006	1,793	1,711	1,661	1,673	n.a.
n.a.	n.a.	3,600	n.a.	n.a.	n.a.
3,083	3,162	3,381	——3,250 per yr, average——		
304 per yr, average——			n.a.	n.a.	n.a.
2,571	2,655	2,691	2,686	2,622	n.a.
1,669	1,710	1,769	1,781	n.a.	n.a.
1,220	1,255	1,273	1,291	1,301	1,288
441	442	487	471	494	508

Hole per bit, development: 1965: *ENP,* 1970:9, p. 4; 1968: *ENP,* 1970:4, p. 16; 1969: *ENP,* 1970:9, p. 4; 1971–72: *ENP,* 1973:9, p. 6.

Cost of wells: All data are "estimate cost." Actual cost is somewhat higher—e.g., for 1965 actual cost was 160 rubles per meter compared with the cost of 148.7 rubles per meter shown in the table (Mel'nikov, 1968, p. 287).

Cost of wells, exploratory: 1965–70: *ENP,* 1971:12, p. 17; 1971: *ENP,* 1973:8, p. 20; 1972: *ENP,* 1973:9, p. 6 (same source says estimate cost increased by 6.15 rubles from 1971 to 1972).

Cost of wells, development: 1965, 1970: *ENP,* 1971:10, p. 7; 1967: *NKh,* 1968:6, p. 4; 1971–72: same sources as for exploratory.

Number of wells completed, exploratory: 1965: Egorov, 1970, p. 31; 1966–70: VNIGRI, 1973b, p. 63; 1971 and 1972: extrapolated by exploratory meterage and average depth.

Number of wells completed, development: 1965: Egorov, 1970, p. 31; 1970: *NKh,* 1971:3, p. 6.

Number of new wells put into production, oil: 1965–70: *ENP,* 1968:9, p. 29 and 1972:6, p. 38; 1971–73: *EG,* 1975:3.

Number of new wells put into production, gas: GP, 1971:3, p. 2.

Average depth of wells, exploratory: 1965: *ESOG,* p. 105; 1966–72: *NKh,* 1974:4, p. 15. These figures are probably slightly different in concept from those

1974, p. 28), and represents a rise in wells of this depth from 18 to 29.1 percent of total exploratory meterage.

More and more wells are being drilled below 4,500 meters (about 15,000 feet). The meterage in this class rose from 4,800 (that is, 1 well) in 1955 to 460,000 in 1971 or about 4 percent of the total (*NKh*, 1972:10, p. 2).

Increasing depth is an important part of the explanation for poor rig productivity and high average cost per meter of drilling. At the same time, one is impressed at how little of this deep drilling the Soviet industry is actually doing—far less, in fact, than the planners had expected. It was expected that by 1975 a significant part of the exploratory meterage would be in the 4,500+-meter class (*NKh*, 1966: 5, p. 1) and that 70 percent of all exploratory drilling would be in the 3,500+-meter class. This is clearly not going to happen. In addition, many of these deep wells are abandoned, for technical or geological reasons, at considerable depths. Of eighteen wells in the 4,500+-meter class drilled in the Chechen-Ingush ASSR in 1965–70, fifteen were never completed or were dry holes (*GNIG*, 1972:10, p. 60).

It seems quite probable that the effort to find and produce new reserves at great depths in some of the traditional areas is seriously hampered by the ineffectiveness of Soviet equipment in drilling wells below about 3,500 meters. This is a serious problem, since it is estimated that about a third of Soviet "predicted" (*prognoznye*) reserves of oil lie at depths greater than 3,000 meters (Feigin, 1974, p. 34).

Notes to table 5, continued

in *ESOG*, referring only to completed wells, and this is somewhat larger than average depth of all exploratory wells.

Average depth of wells, development: 1965: *ESOG*, p. 105; 1966: *NKh*, 1968:11, p. 6 (which gives 2269 for exploratory); 1967: *NKh*, 1969:7, p. 5; 1968: *NKh*, 1970:4, p. 16; 1969: *ENP*, 1970:9, p. 5; 1970: *ENP*, 1971:10, p. 7; 1971: *NKh*, 1973:1, p. 22.

Number of rigs operating: Calculated as total meters drilled, divided by commercial speed, divided by 12. Since not all rig time is included in figuring commercial speed, the total number of rigs on hand is actually appreciably larger than the numbers shown. For example, Shashin cited a figure of 1,414 rigs active in exploration in 1968 compared with the 1,220 calculated in the table.

[a] Of the total 13.6 million meters drilled in 1973, the Ministry of the Oil Industry accounted for about 10.5 million meters, the Ministry of the Gas Industry for about 0.97 million meters, and the Ministry of Geology for the rest.

[b] Of this total only 84 thousand meters with bits 7″ and smaller (Umanskii and Umanskii, 1974, p. 87).

[c] Preliminary reports for 1974 show rig productivity in exploratory drilling as virtually unchanged from 1973, and up about 6 percent in development drilling (*Organizatsiia i upravlenie neftianoi promyshlennosti*, 1975:1, p. 4).

Time Budgets

Weaknesses exist in all elements of drilling technology—quality of bits and drill pipe, mud technology, and in the organization and supply of drilling operations. One of the standard performance indicators in Soviet drilling is a time budget in which the calendar time of an active rig is allocated among the various activities—actually drilling, lowering and raising tools, time lost in accidents, and the like. In connection with the commercial speed indicator, these budgets can be converted to a form showing the expenditure of rig time in various activities per meter drilled, as in table 6. In exploratory drilling, time expenditure per meter has risen since 1965 under every category of the time budget, though most of the increased expenditure has been in the "productive" categories. Through 1972 time spent on stoppages, repair, and other such organizational and qualitative aspects, remained about the same but nothing was happening to offset the greater expenditure of actual drilling time and time spent raising and lowering tools caused by the greater average depth of exploratory wells.

It is a pity that I could not find more detailed recent time budgets for development drilling. But in general discussions by Soviet authors there

Table 6. Time Budgets for Drilling, 1965–74
(minutes per meter drilled)

	Exploratory drilling[a]			Development drilling		
Activity	1965	1970	1972	1965	1970	1974
			Productive time			
Rotating bit	14.6	18.6	19.2	6.5	n.a.	n.a.
Raising and lowering tools	15.3	20.5	21.0	8.3	n.a.	n.a.
Running and setting casing	3.2	8.1	10.0	3.4	n.a.	n.a.
Auxiliary work	30.1	32.3	31.2	9.3	9.8	7.4
Subtotal	63.2	79.5	81.4	27.5	27.1	20.7
			Nonproductive time			
Stoppages for repair	7.9	9.5	5.2	2.3	n.a.	n.a.
Geological complications	3.0	21.2 (both rows)	21.7 (both rows)	2.9	n.a.	n.a.
Breakdowns	15.7			2.1	n.a.	n.a.
Organizational stoppages	24.3	18.1	23.0	4.7	4.8	4.2
Subtotal	50.8	48.7	49.9	12.1	7.6	6.1
Total	114.0	128.2	131.3	39.6	38.6	29.8

Notes: n.a. = not available. Some components may not add to totals because of rounding.

Sources: Exploratory drilling: 1965: Egorov, 1970, p. 34; 1970–72: *ENP,* 1969:1, p. 4; *NKh,* 1974:4, p. 15; *ENP,* 1973:9, p. 6.

Development drilling: 1965: Egorov, 1970, p. 34; 1970 and 1974: *NKh,* 1975:8, p. 3.

[a] 1975 plan total is 74.5 minutes.

is agreement that the reduction in total time spent per meter (from 39.6 to 29.8 minutes between 1965 and 1974) is attributable primarily to the introduction of higher strength drill pipe, more durable bits, and more use of additives to drilling muds (*PKh*, 1971:11, p. 5).

MATERIAL INPUTS TO WELL CONSTRUCTION

Little progress has been made in reducing material inputs to drilling. It was hoped that it would be possible to increase the role of slimhole wells, introduce drill stem testing more widely, and use more light-weight alloy tubular goods for drill pipe and casing. All of these innovations would have helped to reduce materials requirements, had they been widely introduced. As table 7 shows, expenditures of casing, drill pipe, and cement have all increased since the mid-sixties much faster than meters drilled. By 1972 steel used for casing and drill pipe was approaching a million tons per year. It is not clear why the expenditure rates per meter drilled should have increased, but in the absence of simplification of well design they could certainly not come down.

THE TURBODRILL AND ITS REPLACEMENTS

As explained in *ESOG*, improvements in drilling productivity were gained in the past primarily by the introduction of the turbodrill. Under present conditions, however, the original turbodrill has reached the limit of its potential. Below 3,000 meters, in soft rocks, in conditions demanding heavy use of weighting agents in drilling mud, it is not a very effective device.

TABLE 7. AMOUNT AND EXPENDITURES OF MAJOR MATERIALS IN WELL CONSTRUCTION

Material	1940	1955	1963	1972
Kilograms used per meter				
Drill pipe	13.6	10.0	7.9	11.0
Casing	71.3	50.4	43.6	54.1
Cement	48.7	43.6	38.0	49.6
Weighting agents	n.a.	n.a.	n.a.	146.0
Total meterage drilled (1,000 m)	1,947	5,012	9,148	12,720
Total expenditures (1,000 tons)				
Drill pipe	26.6	50.1	72.3	139.9
Casing	138.8	252.6	398.8	688.2
Cement	94.6	218.5	347.6	630.9
Weighting agents	n.a.	n.a.	n.a.	1,857

n.a. = not available.
Source: ESOG, p. 98; Umanskii and Umanskii, 1974, p. 363.

The original attraction of the turbodrill to the Russians was that it enabled them to sidestep many of the problems with quality of pipe and tool joints that made improvement of rotary drilling very difficult. As compared with the rotary drill, the hydraulic transmission system of the turbodrill greatly reduced the stresses and strains on the drill string, even while the amount of power being applied was increasing. It also greatly reduced the amount of time lost in breakdowns, and permitted a large rise in mechanical speed—the rate at which the bit penetrated the rock. Between the late forties and 1955, drilling was shifted almost completely from rotary to turbodrilling, the share of the latter rising to about 85 percent of the total. By 1958 this almost complete changeover to turbodrilling raised rig productivity by 75 percent in development drilling and more than doubled it in exploratory drilling compared with 1950.

But the turbodrill has several disadvantages. Efficient use of the power of the pumps requires a high rpm of the turbine and bit which shortens bit life (measured as meters drilled per bit) and leads to large time expenditures for raising and lowering the drill string to replace bits. The deeper the well, the greater this waste becomes. Furthermore, in deep wells the turbodrill is an ineffective way of getting power to the bottom of the hole because of friction losses. Any effort to move toward slimhole design also undercuts the effectiveness of turbodrilling, and the turbodrill is not suitable for use with jet bits. Finally its combination of high speed, low torque, and light weight on the bit is suitable for the hard rocks of the Volga–Ural region, but less so for the softer rocks of other regions. By the mid-sixties, Soviet drilling experts were well aware of these disadvantages and sought a variety of solutions: redesign of the turbodrill itself, experimentation with electric drills, and lengthening bit life through improving the quality of tricone bits or replacing them with diamond bits.

So far, only modest progress has been made along any of these lines. As table 5 shows, average bit life in 1965–72 rose only from 17 to 20 meters in exploratory drilling, but from 28.3 to 39.6 meters in development drilling. Experimentation with diamond bits has shown that their much longer life (that is, about 56 meters in exploratory drilling in 1970 versus 20 meters for tricone bits) provides large savings in bits and rig time. But there seems to be some obstacle to the extension of this apparently successful experiment, since meterage drilled with diamond bits (281,000 meters in 1972) is a very small share of the total and seems not to be increasing. Policy statements recently have stressed the introduction of improved tricone bits (mentioned in the Ninth Five Year Plan as a high-priority innovation) more than diamond bits, and it may be that diamond bits will remain small in total drilling. According to *EG*, 1974:16, p. 5, the Ministry of the Oil Industry in the first quarter of 1974 drilled 300,000 meters with bits made of special treated steel. These bits are supposed to have a service life 10 to 15 percent greater than traditional bits, but their share in the total is still too small to affect average bit life.

There was for a while some inclination to shift back from turbodrilling to rotary drilling, but oil industry officials have either changed their minds or have been unable to carry out this intention. The share of turbodrilling dropped from about 85 percent in the early sixties to about 74 percent of the total by 1970, but now seems to be stabilized at that figure. Turbodrills of modified design have been developed to give the desired effects of lower speed and higher torque. The modifications include use of multiple-section, variable-line-of-pressure design, better bearings, and the use of a reduction gear. It is my impression, however, that, as of the mid-seventies, all of these new designs are still in the experimental stage and are not yet being produced and used in the industry's drilling program.

Another potential replacement for the turbodrill is the electric drill. But it seems that the many problems with this technology, described in *ESOG*, have never really been overcome. The share of electric drilling has stayed more or less constant—at about 2 percent since 1965. At the height of experimentation with electric drills in the mid-sixties, several hundred a year were being produced (220 in 1965), but output has fallen steadily since then to only 112 drills in 1973. A few are being imported from France. It seems likely that either the factories that might produce the electric drill have no interest in producing it or in making the modifications needed to improve it, or that there is some hesitation on the part of oil industry officials about extensive introduction of the electric drill on the basis of experience so far.

Table 5 shows several other important indicators, such as the number of wells completed and the number of rigs operating. Recently, Soviet sources have not reported the number of wells completed, but the numbers can be figured approximately by comparing total meterage and average depth. (Some of the meters drilled are lost in wells abandoned before reaching the intended depth—about 15 percent in exploratory wells and about 12.5 percent in development wells in 1965.) This calculation suggests that the number of exploratory wells drilled per year fell from 2,165 in 1965 to about 1,600 in 1972, while the number of development wells increased from 2,738 to 3,600 in 1970. Preliminary figures suggest that development wells drilled increased further, to about 4,300 in 1974. This is another indication of the way effort is being focused on expansion of production, perhaps to the neglect of developing new reserves to support future production.

The series on rig productivity in connection with meterage also permits an estimate of the number of rigs operating (also shown in table 5). The number of rigs available to the industry has increased only slowly—10 percent in seven years against much faster growth in the fifties and early sixties. The number increased by three-quarters in the seven years before 1965. These data also show strikingly how low productivity in exploratory drilling mortgages total rig availability—almost three-quarters of the 1,795 rigs active in 1973 were committed to exploratory drilling.

Testing

Well testing remains a serious problem, with long delays and wastage of time and materials. Average time spent testing a well, after drilling is completed, is three months (more for exploratory wells, less for development wells). But only about a third of this is actually spent in testing; the rest is dead time caused by waiting for crews and equipment, organizational delays, and so on (*GNIG*, 1973:2, p. 16; and Feitel'man, 1969, p. 144). The fraction of wells tested with drill-stem testers is still small. The Ministry of the Oil Industry has increased the share of objects tested with drill-stem testers up to about one-fourth of the total (*Organizatsiia i upravlenie neftedobyvaiushchei promyshlennosti*, 1970:7, p. 13), but the Ministry of Geology has done less well. Most dry holes (about 80–85 percent) have been cased (Feitel'man, 1969, p. 144), which means large expenditures of metal to no useful purpose. To illustrate: in the process of drilling, the drill crew gets a show of oil or gas in the mud returned from the bottom of the hole, but lacking the kind of equipment that would permit them to test the well without first casing it, they proceed to drill it to the specified depth, then case and test it. By then the permeability of the formation may have been so reduced as to cause a misleading test or to interfere with the productivity of the well.

Also, testing only after casing means interfering with the flow of oil from the formation. Drillers get a show but, having no tester, go ahead with drilling. They then case the well and test it, and get no indications of oil.

Offshore Operations

One of the most urgent tasks facing the Soviet oil and gas industry is to master the technology of offshore drilling. The total Soviet shelf area is said to be about 8 million km², of which 2.5 million km² is promising for the discovery of oil and gas (*Neftepererabotka i neftekhimiia,* 1970: 8, p. 40). Especially attractive are the Arctic, Baltic, Black, and Okhotsk Seas, since production here would create no location problem. The obstacle is the lack of equipment needed for offshore drilling. The only significant offshore experience so far has been in the Caspian Sea, where significant amounts have been drilled (see table 8).

The Soviet approach has generally been to use fixed platforms, artificial islands, and trestles, rather than mobile rigs. But the investment requirements for this approach are high, and it is very risky without strong assurance that oil will be found. A shift to mobile platforms is intended, but so far little has been accomplished. (The Turkmen figure essentially represents the amount of drilling that has been done from mobile rigs.) The USSR has built two prototypes of mobile offshore rigs. The "Apsheron" is limited to water depths not exceeding 15 meters, and is capable of drilling only to 1,800 meters. The "Azerbaidzhan" is a heavier-duty rig that can work in depths to 20 meters, and can drill to 3,000 meters. To get into deeper water (up to 60 meters), and to drill to the 4,000-meter depths where it is expected oil will be found in the Caspian area, a rig called the "Khazar" was purchased in the Netherlands. This has now been working for some time in the Turkmen side of the Caspian. The Russians planned to produce domestically a series of ten units of this type, to be called the "Caspian." As of 1974, however, none had been completed and put into operation (*OGJ*, June 11, 1973, p. 71); indeed the Russians are still describing this as a "project" (*NKh*, 1974:7).

Up to this writing, the only Soviet offshore effort in the Arctic has been on Kolguev Island in the Barents Sea. In the Black Sea a fixed

TABLE 8. OFFSHORE DRILLING IN THE CASPIAN SEA, 1960–73

(thousand meters)

Year	Total	Azerbaidzhan[a]	Turkmen
1960	137.9	61.9	None
1965	[b]	88.2	3.5
1966	[b]	[b]	5.0
1968	[b]	[b]	0.4
1970	200	134.6	10.6
1973	[b]	[b]	30.2

Sources: Agaeva, Baku, 1972, pp. 40, 82; and *GNIG,* 1974:1, p. 6.

[a] Total exploratory footage for Azerbaidzhan in 1966–70 is reported in another source as 609,700 meters (VNIGRI, 1974, p. 93), which suggests that the figures tabulated may be incomplete.

[b] The unaccounted-for meterage is probably in the Chechen-Ingush sector of the Caspian.

platform has been constructed. No wells have yet been drilled offshore in the Baltic or Okhotsk Seas, though one of the Ninth Five Year Plan goals was to begin drilling offshore near Sakhalin (*PKh*, 1972:11, p. 36). Apparently this was to be done through slant wells from onshore locations (VNIGRI, 1972, p. 91). The Minister of the Oil Industry recently asserted the urgency of doing preliminary work now to make possible the drilling of 60 to 70 thousand meters a year in the Okhotsk Sea after 1975 (*NKh,* 1974:3, p. 4). Finally, the Russians have an agreement with a consortium of Japanese oil companies to drill exploratory wells on the continental shelf off Sakhalin, using rigs provided by the Japanese. As of 1975, however, the prospect of actual drilling was still some time in the future (Moscow Narodny Bank, *Press Bulletin,* No. 709).

4

OIL PRODUCTION

Oil output has continued to grow rapidly, as shown in table 9, though at a somewhat diminished rate of increase in the seventies compared with the sixties.[1] The dominant consideration now affecting most indicators of production performance is depletion of the large fields in the Volga–Ural area that were the key to output expansion up to about 1970. This has led to a rise in cost, a shift in regional output distribution, and a concern for new production methods.

The Problem of Depletion

As reservoir pressure has been depleted, the share of flowing wells in total output has fallen from 64.4 percent in 1965 to 40 percent in 1973 (see table 10). More and more wells must be transferred to some kind of mechanized production technique, which will raise production costs. The standard pumping jack is inappropriate for Soviet waterflood operations. In the postwar period all the big Soviet fields have been designed for production with pressure maintenance through waterflooding, and as the injected water reaches the production wells it becomes necessary to extract very large volumes of liquid if considerable amounts of oil are to be recovered. This is done with submersible electric pumps, and wells so equipped now account for over one-third of all oil output. Until recently most of these pumps were of Soviet domestic manufacture. However, because the Soviet pumps are unreliable and unproductive, submersible electric pumps have become a major item in Soviet machinery imports from the United States.

Another consequence of depletion is reduction in output per well. Average well productivity for the USSR as a whole has actually risen,

[1] Table 9 differs from the comparable table in the Appendix in showing condensate separately from fuel oil. In recent years the production of crude oil in the USSR has begun to be supplemented with significant amounts of condensate produced jointly with natural gas. In the past, the Russians lumped condensate output together with crude oil under the general heading of oil output. They now report condensate output separately, though statements about the output of oil (*neft'*) in general, still include condensate.

TABLE 9. OUTPUT OF CRUDE OIL AND CONDENSATE

(million tons)

Year	Crude oil including condensate	Condensate	Crude oil
1959	129.6	0.4	129.2
1960	147.9	0.7	147.2
1961	166.1	0.7	165.4
1962	186.2	0.8	185.5
1963	206.1	0.8	205.2
1964	223.6	1.0	222.6
1965	242.9	1.2	241.7
1966	265.1	1.4	263.7
1967	288.1	n.a.	n.a.
1968	309.2	1.7	307.4
1969	238.4	2.3	326.1
1970	353.0	4.2	349.3
1971	377.1	5.3	371.8
1972	400.4	6.6	393.8
1973	429.0	7.6	421.4
1974	458.8	6.8	452.0
1975 (Plan)	489.4	7.4	482.0
1975 (9th FYP)	505.0	9.0	496.0

Notes: n.a. = not available. In some cases components do not add to totals because of rounding.

Sources: There is apparently some ambiguity in the definition of condensate output. Baibakov, 1972, reported 1970 output as 3.8 million tons, but when condensate output began to be shown in Central Statistical Administration handbooks, the 1970 figure was given as 4.2 million tons. Gas industry sources are citing output figures that are much smaller than the handbook sources—e.g., 5.9 million tons for 1973 (*GP,* 1974:1, p. 1) as compared with the 7.6 reported in handbooks. Since there seems no possibility of condensate output outside the gas industry, it may be that the Central Statistical Administration is including some other kind of liquid hydrocarbon in its total.

Crude oil including condensate: 1959–73: *Nar khoz,* various years; 1974: *EG,* 1975:3; 1975 Plan: *EG,* 1974:52; 1975, 9th FYP: Baibakov, 1972.

Condensate: 1959–65: Mel'nikov (ed.), 1968, p. 461; 1966: Feigin, 1974, p. 7; 1968–69: Lisichkin (ed.), 1974, p. 68; 1971–73: *Nar khoz,* 1973; 1974: *EG,* 1975:3; 1974 Plan: by subtraction; 1975, 9th FYP: Baibakov, 1972.

Crude oil: By subtraction, except 1975 Plan, from *ENP,* 1975:7, p. 30.

because output per well for new wells is quite high, especially for Western Siberia (2,656 tons per well per month in 1972). A rising share of these new areas in the total keep the all-union average up. But in the big fields of the Volga–Ural region, output per well per month has fallen sharply since 1965—by 51 percent in Bashkir ASSR, 24 percent in Tatar ASSR, and 17 percent in Kuibyshev oblast'.

Depletion also means an increasing gap between the amount of new capacity developed and the net increment in output as more and more new capacity goes to replace output declines from existing wells and

TABLE 10. SELECTED INDICATORS RELATING TO PRODUCTION METHODS

Item	1965	1966	1967
Crude oil output (mill. tons[a])	241,688	263,724	286,707
By type of lift (percent)			
Flowing wells	64.4	61.3	58.9
Pumped wells	33.5	36.7	39.0
Electric submersibles	14.9	n.a.	n.a.
Gas lift	1.8	1.8	1.8
Other	0.3	0.2	0.3
Output from pressure maintenance fields (percent of total)	70	n.a.	70
Water injected (million m³)	328.9	368.7	426.1
Number of hydrafrac treatments	2,083	2,130	2,166
Output per well-month (tons)	509.8	530.0	548.9
New wells	1,062	n.a.	1,170
Total well-months (thousands)	474	498	522
Implied average number of active oil wells	44,272	46,512	48,206
Number of oil wells (Jan. 1)	n.a.	51,058	n.a.
Active	42,270	46,313	n.a.
Producing	n.a.	40,755	n.a.
Inactive and awaiting commissioning	n.a.	4,745	n.a.
Output divided between[c]			
Old wells (mill. tons)	226.5	247.3	267.9
New wells (mill. tons)	14.2	15.6	17.5
Decline in wells carried over (mill. tons)	n.a.	16.5	21.2

n.a. = not available.

Sources: Crude oil production: from table 9.

Production by type of lift: 1965–68: *Nar khoz,* 1968, p. 235, except electric pumps—*NKh,* 1972:1, p. 53, and Egorov, 1970, p. 230; 1970: *NKh,* 1972:1, p. 53, 1975:3, p. 24, and 1974:7, pp. 28, 80, *PKh,* 1973:4, p. 16; 1971: *NKh,* 1973:1, pp. 23–24; 1972: Umanskii and Umanskii, 1974, p. 100; 1973, 1975 Plan: *NKh,* 1974:7, pp. 28, 30.

Output from pressure maintenance fields: Standard series in statistical handbooks, except 1975 Plan—*NKh,* 1973:12, p. 28.

Water injected: Standard series in statistical handbooks.

Hydrafrac treatment: Standard series in statistical handbooks.

Output per well-month: All wells—Shmatov, *et al.,* 1974, p. 41, except 1975 Plan, from *NKh,* 1972:5, p. 2; new wells—Brenner, 1968, p. 126, and *ENP,* 1968:9, p. 29.

Total well months: Calculated from output, and output per well-month.

Implied number of active oil wells: Well months times 30 divided by 365 gives average annual number working. But wells working were only 88 percent of active stock on Jan. 1, 1966, 91.2 percent on Jan. 1, 1970, 96.6 percent on Jan. 1, 1971, 96.6 percent on Jan. 1, 1972 (Shmatov, *et al.,* 1974, p. 40). So the last step is to adjust upward using these coefficients.

1968	1969	1970	1971	1972	1973	Plan 1975
307,448	326,083	349,256[b]	371,776	393,840	421,400	496
57.7	n.a.	55.1	n.a.	50.2	48.0	n.a.
40.3	n.a.	44.1	45.6	47.4	48.7	n.a.
20.2	n.a.	23.8	26.2	29.2	34.7	33.3
1.8	n.a.	n.a.	n.a.	2.4	3.3	4.0
0.2	n.a.	n.a.	n.a.	0.0	0.0	n.a.
n.a.	72	73	73	75	79	75
478.9	512.1	559.0	618.2	691.5	794.7	n.a.
1,672	1,396	n.a.	n.a.	n.a.	n.a.	n.a.
562.8	579.7	589.9	601.3	613.1	n.a.	664
n.a.	n.a.	n.a.	n.a.	n.a.	n.a.	n.a.
546	563	592	618	642	n.a.	749
50,141	51,415	53,765	54,652	54,624	n.a.	63,745
n.a.	n.a.	53,540	53,488	57,500	59,574	n.a.
50,000	n.a.	50,350	50,760	54,855	n.a.	n.a.
n.a.	n.a.	n.a.	50,557	53,000	57,544	n.a.
n.a.	n.a.	3,190	2,728	4,500	n.a.	n.a.
288.0	305.7	326.4	345.5	n.a.	n.a.	n.a.
18.3	19.2	21.2	25.1	n.a.	n.a.	n.a.
25.3	30.2	32.7	34.8	n.a.	n.a.	n.a.

Number of oil wells: NKh, 1974:7, p. 28. *ENP*, 1973:9, pp. 3–4. In the Soviet classification of oil wells the broadest category is the "exploitation fund," divided into "active," "inactive," and "wells awaiting commissioning." Not all "active" wells are actually producing at a given time. Statistics for these concepts are usually given as of a given date, but occasionally the data cited seem to be based on some kind of average annual concept. There are numerous inconsistencies in the available data, but the figures in the table represent a considered balancing of all data, mainly from the following sources: Shmatov, *et al.*, 1974, p. 40; Umanskii and Umanskii, 1974, p. 25; *NKh*, 1973:12, p. 3, 1974:3, p. 7, 1974:7, p. 28; and *ENP*, 1973:9, p. 3, 1974:7, p. 21.

Output divided between old and new wells: One ambiguity in this breakdown is whether "new wells" includes only newly drilled wells or also wells returned from capital repair; both concepts are used but the data here are for the first (*ENP*, 1973:7, p. 23). Output from the two kinds of wells is slightly less than total output because some oil is produced outside the ministry, to which these data are limited. Decline in wells carried over—*ENP*, 1973:7, p. 23.

[a] Differs from output as shown in other tables because condensate is excluded here.

[b] Components exceed total because of inconsistency in sources.

[c] Differs from line 1 because it covers Ministry of Oil Industry only.

reservoirs. The general picture is shown by the estimates below, based on statements by N. K. Baibakov in *NKh*, 1974:7, p. 2 and Luzin, 1974, pp. 4, 22, 79.

	Gross additions to production capacity	Decline in output in old fields		Net increment in production capacity
	(million tons)	(million tons)	(% of gross)	(million tons)
1961–65	154	60	39	94
1966–70	223	115	52	108
1971–75	407	261	64	147

If these recent statements are compared with some earlier ones, it appears that output declines in old fields have been much greater than expected. Total new capacity required to achieve planned growth in 1971–75 was earlier given at various times as 300, 310, 320, and 384 million tons; assertions as to the absolute amounts and shares of the decline from existing fields have also involved several upward revisions.[2] The unforeseen magnitude of replacement needs helps to explain current concern about the adequacy of reserves and drilling activity and the urgent efforts at secondary recovery. I suspect that the Soviet oil industry has been meeting output goals by intensifying production from existing fields still further, in part by drilling additional production wells beyond the number originally envisaged for particular reservoirs. This would both accelerate the exhaustion of existing fields and hinder the growth of new reserves, since exploratory drilling is being sacrificed. This situation will no doubt get worse rather rapidly. The pattern for pressure maintenance projects is for output to be more or less steady for eight to ten years; but when it begins to fall, it is expected to do so quite rapidly.

The depletion problem has stimulated a frantic effort to improve methods of secondary recovery. Soviet oil industry officials are eloquent about the advantages of raising recovery rates. The location of this oil is already known, it is close to markets, and is already provided with production, transport, and refinery facilities. Secondary recovery thus holds out the hope of avoiding some expensive investments in finding and developing new reserves. Projects and experiments are being carried out with all known methods—using surfactants, increasing the viscosity of flood water to improve flushing, using solvents such as natural gas liquids, repressuring with air and dry gas, and thermal methods. But

[2] There are problems of interpretation here, since some statements refer to amounts needed to cover drop in output from old *wells*, others from old *fields*, and output capacity may be figured on an average annual basis or on an end-of-year basis.

apparently this effort is not conducted on a very large scale. N. K. Baibakov (formerly Minister of the Oil Industry and now chairman of Gosplan), has been very insistent on the virtues of such a strategy, and he has expressed disappointment that, in 1973, only 1.3 million tons of oil were produced as the result of all the secondary recovery methods being used (*NKh*, 1974:7).

Secondary recovery is likely to face serious institutional obstacles in the USSR, analogous to those described in *ESOG* for hydraulic fracturing. According to one Soviet official, most of the methods described above involve heavy costs in the early stages, which will raise output only later. Oil field administrators are thus likely to be skeptical and reluctant to use such methods since they worsen the performance indicators on which bonuses depend, even though such expenditures are economically justified from a branch-wide, or a long-term perspective.

Depletion in the old areas has meant a significant shift in the regional distribution of oil production, as shown in table 11. The Volga–Ural region dropped from 71 percent in 1965 to 46 percent in the plan for 1975. Azerbaidzhan's output has declined in absolute terms and is now insignificant in the total picture, with only 4 percent of total output in 1972. The North Caucasus has had enough new discoveries to increase its output and hold its share, as has the Ukraine. The major source of new output has been West Siberia, where output was virtually nothing in 1960, but was expected to reach 146 million tons in 1975, which would be almost 30 percent of all Soviet oil output in that year. This is in fact considerably greater growth than planned when the Ninth Five Year Plan was approved (it set a target of only 125 million tons). This extra performance has been necessary to offset the unexpectedly large declines in output from other fields in old regions and the disappointing growth performance of some other new areas. For example, there were also great hopes for Western Kazakhstan, but it has failed to meet expectations, and produced only 5 percent of total Soviet output in 1973. Several other new areas—Komi ASSR, Belorussia, and the Turkmen SSR—have raised output quite rapidly, but are still relatively small contributors to the total.

Cost of Production

The production cost for oil, as the Russians figure it, has risen from about 3 rubles per ton in 1965 to about 4.37 rubles per ton in 1971, or by almost 50 percent (see table 12). This is partly the result of the price reform in 1967, which raised prices on some of the inputs used in oil production and also introduced a charge for finding costs. It is said

TABLE 11. SOVIET CRUDE OIL OUTPUT, BY REGIONS, 1965–73, AND 1975 PLAN[a]
(million tons)

Region	1965	1966	1967	1968
RSFSR	199.9	218.0	235.0	251.5
Volga–Ural	173.6	185.7	196.9	n.a.
Tatar ASSR	76.4	83.1	89.0	95.9
Bashkir ASSR	43.9	46.6	48.1	n.a.
Kuibyshev oblast'	33.4	33.5	34.0	n.a.
Volgograd oblast'	6.2	n.a.	6.3	n.a.
Saratov oblast'	1.3	1.2	1.2	n.a.
Orenburg oblast'	2.6	3.5	4.3	n.a.
Perm oblast'	9.7	11.5	13.0	14.6
Udmurt ASSR	0.0	0.0	0.0	0.0
Western Siberia	0.9	2.8	5.8	12.2
North Caucasus	20.0	22.9	25.9	n.a.
Krasnodar krai	5.4	5.3	5.5	n.a.
Chechen-Ingush ASSR	9.0	11.2	13.5	16.2
Dagestan ASSR	1.0	1.2	1.4	n.a.
Stavropol' krai	4.5	5.1	5.5	n.a.
Sakhalin	2.4	2.6	2.6	n.a.
Komi ASSR	2.2	3.1	3.8	4.4
Belorussian SSR	[b]	0.2	0.8	1.7
Azerbaidzhan SSR	21.5	21.7	21.6	21.1
Offshore	n.a.	n.a.	11.0	n.a.
Ukrainian SSR	7.6	9.3	11.0	12.1
Central Asia	11.8	12.8	14.1	15.2
Turkmen SSR	9.6	10.7	11.9	12.9
Uzbek SSR	1.8	1.7	1.8	1.8
Kirgiz SSR	0.3	0.3	0.3	0.3
Tadzhik SSR	[b]	0.1	0.1	0.1
Kazakh SSR	2.0	3.1	5.6	7.4
Georgian SSR	[b]	[b]	[b]	[b]
All USSR[c]	242.9	265.1	288.1	309.2

Notes: n.a. = not available. The regional distribution given here, based on later sources, differs slightly from that for 1965 in *ESOG*.

Sources: All data from standard statistical handbooks, except as noted. There are many inconsistencies in the sources, and data have sometimes been adjusted to reconcile them with regional subtotals.

1965—Volga–Ural, North Caucasus, Krasnodar, Dagestan, Stavropol': Mingareev and Luzin, 1972, p. 33; Tatar, Bashkir, Kuibyshev, Volgograd, Saratov, Perm, Chechen-Ingush, and Sakhalin: *EG,* 1966:9, pp. 10–11; Orenburg: *NKh,* 1971:3, p. 23.

1966—Volga–Ural, North Caucasus, Sakhalin, Tatar, Bashkir, Kuibyshev, Saratov, Perm, Krasnodar, Chechen-Ingush, Dagestan: Brenner, 1968, pp. 33, 178; Orenburg: by handbook index over 1965; Western Siberia: *NKh,* 1972:12, p. 2; Komi: *ENP,* 1971:11, p. 3.

1967—Most of the detail is from Egorov, 1970, p. 76; Perm: *GNIG,* 1968:1, p. 1; Komi: *ENP,* 1971:11, p. 3; Tatar and Bashkir: *GNIG,* 1968:5, p. 1.

1968—Western Siberia: *ENP,* 1970:4, p. 7; Komi: *ENP,* 1971:11, p. 3.

1969—Volga–Ural, Western Siberia, North Caucasus, Krasnodar, Dagestan and Stavropol': Mingareev and Luzin, 1972, pp. 33, 40; Komi: *ENP,* 1971:11, p. 3.

1969	1970	1971	1972	1973	1975 Plan (excluding condensate)
265.7	284.8	304.4	325.6	351.0	398.3
203.8	208.4	211.2	215.5	n.a.	226.6
98.8	100.3	102.6	102.1	103.0	101.0
n.a.	40.7	40.0	40.1	n.a.	40.0
n.a.	35.0	35.4	35.5	n.a.	35.0
n.a.	7.0	n.a.	} 7.0	n.a.	n.a.
n.a.	1.4	n.a.			
n.a.	7.4	n.a.	9.4	n.a.	14.0
15.3	16.1	16.9	17.8	n.a.	21.5
0.0	0.5	n.a.	1.3	n.a.	5.0
21.8	31.4	44.8	62.7	88.7	125.0
32.6	35.0	35.8	34.6	n.a.	34.1
5.9	6.0	n.a.	5.5	n.a.	n.a.
18.5	20.3	21.6	19.9	n.a.	n.a.
2.1	2.2	2.1	2.0	n.a.	n.a.
6.1	6.5	n.a.	6.9	n.a.	n.a.
2.7	2.5	n.a.	2.4	n.a.	2.6
4.7	7.6	9.1	10.6	n.a.	10.0
2.8	4.2	5.3	5.8	7.0	8.5
20.4	20.2	19.2	18.4	18.2	18.8
n.a.	12.9	n.a.	11.8	11.9	n.a.
13.4	13.9	14.3	14.5	14.1	16.4
16.0	16.8	17.8	18.0	18.2	n.a.
13.7	14.5	15.5	16.0	16.2	22.0
1.8	1.8	1.8	1.6	1.5	1.8
0.3	0.3	0.3	0.3	0.2	0.4
0.1	0.2	0.2	0.2	0.2	0.2
10.2	13.2	16.0	18.1	20.4	30.0
[b]	[b]	[b]	[b]	[b]	n.a.
328.3	353.0	377.1	400.4	429.0	496.0

1970—Volga–Ural, Western Siberia, North Caucasus, Bashkir, Krasnodar, Volgograd, Saratov: Mingareev and Luzin, 1972, pp. 33, 36, 40; Kuibyshev, Orenburg, Perm: *NKh,* 1971:12, p. 23; Chechen-Ingush: *NKh,* 1972:12, p. 14; Sakhalin: *EG,* 1973:5, p. 2; Udmurt: Umanskii and Umanskii, 1974, p. 177.

1971—Volga–Ural: N. A. Eremenko *et al.,* in *AAPG Bulletin,* September 1972, p. 1713; Tatar, Kuibyshev, Perm, Chechen-Ingush: *NKh,* 1972:12, p. 12; Bashkir: *PKh,* 1973:4, p. 16; North Caucasus: Eremenko, op. cit., p. 1714; Western Siberia: *EG,* 1974:27, p. 1; Komi: Anufriev, 1973, p. 5.

1972—All intrarepublic detail from *NKh,* 1973:12, p. 3; *EG,* 1973:5, p. 2; and *GP,* 1973:9, p. 23.

1973—All intrarepublic detail from *NKh,* 1974:3, p. 2, 1974:4, p. 63 and *GNIG,* 1974:3, pp. 1, 2.

1975 Plan—Baibakov, 1972, p. 103; *ENP,* 1973:10, p. 2; Kim, 1973, p. 106; Umanskii and Umanskii, 1974, p. 177.

[a] Including condensate, except as noted otherwise.

[b] Output is negligible.

[c] In some cases components may not add to totals because of rounding.

TABLE 12. COST OF OIL EXTRACTION, BY REGION

(rubles per ton)

Region	1965	1966	1968	1970	1971
USSR average	2.99	3.03	4.17	4.29	4.37
RSFSR average	2.15	2.19	n.a.	n.a.	3.73
Volga–Ural	n.a.	n.a.	n.a.	3.30	n.a.
Tatar ASSR	1.46*	1.55*	2.47	2.70	2.88
Bashkir ASSR	2.48	2.63*	3.89	4.98	n.a.
Kuibyshev oblast'	1.67	1.79*	2.65	2.87	3.00
Saratov oblast'	n.a.	4.90*	n.a.	5.81	n.a.
Perm oblast'	n.a.	1.86*	n.a.	3.73	3.70
Western Siberia	10.30*	n.a.	n.a.	3.54	3.40
North Caucasus	n.a.	n.a.	n.a.	3.82	4.05
Krasnodar krai	n.a.	6.18*	n.a.	n.a.	n.a.
Chechen-Ingush ASSR	2.17	2.17*	n.a.	2.87	2.88
Dagestan ASSR	5.50	4.89*	n.a.	n.a.	n.a.
Komi ASSR	n.a.	n.a.	n.a.	n.a.	6.43
Sakhalin	n.a.	8.43*	n.a.	n.a.	18.00
Belorussian SSR	n.a.	n.a.	n.a.	n.a.	4.02
Azerbaidzhan SSR	7.56	7.70*	10.37*	10.72	11.80
Ukrainian SSR	2.66	2.55*	4.78*	4.85	5.03
Central Asia	n.a.	n.a.	n.a.	5.49	5.72
Turkmen SSR	3.46*	3.33*	5.37*	4.38*	n.a.
Uzbek SSR	8.52	10.13	13.28*	n.a.	n.a.
Kazakh SSR	8.97	8.57	n.a.	5.32	5.42

n.a. = not available.

* = absolute figures, other figures were obtained as indicated below.

Sources: This table is based on three kinds of information; (a) absolute ruble figures for individual regions; (b) indexes of change for regions and the USSR: (c) indexes of cost in different regions compared with the all-union average. Absolute figures for 1966 and 1965 are from Brenner, 1968, p. 178; Kim, 1973, p. 35; the 1966 Tatar ASSR statistical handbook; and *Organizatsiia i upravlenie neftianoi promyshlennosti* 1970:4, p. 33. For 1968 there is a figure for the Ukraine from *Neftianaia i gazovaia promyshlennost',* 1970:2, pp. 8–12, for Azerbaidzhan from Abramov, 1971, and for Turkmen SSR from Kim, 1973. For 1969 Western Siberian cost is given in *Organizatsiia i upravlenie neftianoi promyshlennosti,* 1970:4, p. 35, and Turkmen SSR cost in Kim, 1973. The 1970 figure for Turkmen SSR is also from Kim, 1973.

Cost indexes from Adamesku (ed.), 1973, p. 88, generate figures for the Tatar and Bashkir ASSR's and Kuibyshev oblast' and for the USSR for 1968 and 1970. A fairly complete set of regional relatives for 1970 from Umanskii and Umanskii, 1974, p. 176, supplemented by others in Kozyrev, 1972, p. 50, permits filling in much of that column.

All-union cost for 1965 is figured as 2.99 rubles on the basis of an index concerning cost change from 1960, figured in *ESOG* as 3.30 rubles. Other indexes for 1960–66 imply 3.03 rubles for 1966 (see *ENP,* 1972:6, p. 39).

The all-union figure for 1971 is based on statements in *PKh,* 1973:8, pp. 46–47 and *ENP,* 1973:1, p. 5. Regional figures for 1971 are then calculated by a large set of regional relatives given in *NKh,* 1973:12, p. 4.

The final step is to fill in the 1965 and 1966 columns by a combination of absolute numbers, regional relatives and year-to-year indexes. The main sources are Adamchuk, 1968, p. 188 and *ENP,* 1972:6, p. 39.

that if an adjustment is made to eliminate the effect of these charges, cost rose by only about 10 percent (*ENP,* 1972:6, p. 39).

This, incidentally, calls attention to the conventional nature of Soviet cost accounting for oil production. Though a finding cost has been included as a cost since 1967, the rent charge and capital charge also introduced at that time are still not included in the reported production cost, but are treated as deductions from profit. The aggregate amount of the rent charge in 1970 was 936 million rubles (for oil alone) and the capital charge (figured at 12 percent on the depreciated value of the industry's fixed assets) was about 488 million rubles (Luzin, 1974, p. 65). Spread over the 353 million tons of oil produced in 1970, these charges would raise costs by about 4.00 rubles per ton to a total of 8.37 rubles. This is probably still not a very precise measure of real economic cost, but is a suggestive indication that reported costs are an unreliable measure of the real cost of oil.

It is interesting that the cost of Western Siberian oil is not especially high—at 3.50 rubles per ton at the beginning of the 1970s it is well below the avereage cost for the branch as a whole. Moreover, Western Siberia's share in the industry's total capital stock is appreciably lower than its share in total output, so that more sophisticated cost measures would enhance its cost advantage.

5

OIL TRANSPORT

Since only some features of oil transport are important for the issues to which this study is directed, and since transport involves relatively few controversial policy issues, general developments since 1965 can be summarized rather briefly, and most of our attention can be focused on the cost of transporting oil from Siberia.

The general geography of oil movement involves flows from production centers in the Caucasus, the Volga–Ural region, and now Western Siberia, to the big consuming areas and export points in the west. Important new features include a big pipeline from Samotlor in Tiumen' oblast' to Al'metevsk in the Tatar ASSR, finished in summer 1973, and now in operation. This line is constructed of 1,220 mm (48 inch) pipe and is over 2,000 kilometers long. Construction started on a second such line in summer 1974. It will have the largest throughput capacity of any Soviet oil pipeline. There are also smaller movements of crude oil and products eastward into Eastern Siberia and the Far East, and into Central Asia. The opening of fields in areas which did not at first have access to pipelines meant temporary dependence on railroads for movement of crude, as in Western Siberia, but pipelines were soon built to handle them. River transport and ocean transport haul mostly oil products rather than crude oil.

The main economic issue in transport seems to be how fast the shift to pipelines should be made. Railroad oil traffic has increased greatly—there were increments of 120 million tons in shipments and 130 billion ton-kilometers in turnover (that is, tonnage shipped times the average length of haul) between 1965 and 1972. One suspects that it would have been desirable to have moved still faster in transferring some of this transport load to pipelines. Pipelines are cheaper for refined oil products as well as for crude oil and there are surely many product flows now that would justify the construction of pipelines. Soviet studies, in fact, conclude that product lines should play a much larger role than they now do in product transport (Ushakov, 1972, p. 166). A few product pipelines have been built from the Volga–Ural region both eastward and westward. But pipeline construction capacity is one of the biggest bottlenecks in the oil and gas industry, hindered both by pipe shortages and limited construction capacity. Under these conditions building effort has been allocated predominantly to gas and crude oil lines, where pipelines have the greatest comparative advantage.

The main indicators of Soviet oil transport are shown in table 13. The growth of oil output since 1965 has meant a big increase in the amount of oil to be transported, and locational shifts have increased the average length of haul. Hence, turnover has increased much faster (1.95 times between 1965 and 1973) than the amount of oil shipped (1.7 times). To cope with this additional work, reliance has been placed mostly on pipelines and railroads; the amount of oil transported by river and ocean tankers has increased much less.

The increased scale of many flows and operations—regional outputs, average size of refineries, individual interregional flows—encourage the movement of refining from producing to consuming centers. The regional distribution of oil refining is shown in the tabulation below (Gankin *et al.* [eds.], 1972, p. 102). Moving crude oil from the field to refineries is the task for which pipelines offer the greatest advantage compared with other modes, and there has been a steady decline in the share of the railroads in total turnover in favor of pipelines.

Region	1965	1969
	(percent of total)	
Caucasus	21.3	18.4
Volga–Ural	43.1	38.7
Other regions in European USSR	20.1	26.0
Asiatic regions	15.5	16.9
Total	100.0	100.0

This shift in proportions has been accompanied by a sharper division of function between the two carriers—pipelines for crude and the railroads for products. The share of products in pipeline work is very small (7 percent of turnover in 1971), and the railroads have been relieved somewhat of the job of hauling crude.

The greater-than-planned increase in Western Siberian output has had important consequences for the pipelines; that is, a much faster growth of pipeline transport than targeted in the Ninth Five Year Plan. The annual plan for 1975, approved in 1974, set oil pipeline shipments at 442.2 million tons and oil pipeline turnover at 619.2 billion ton-kilometers, both of which are well above the original planned figures. The above-plan importance of Western Siberian oil has both limited the possibility of using other carriers and has increased the average length of haul to 1,400 kilometers compared with the planned 1,338 kilometers. To handle the extra work, the pipeline network has had to be extended

TABLE 13. SELECTED INDICATORS RELATING TO TRANSPORT OF OIL AND PRODUCTS

Item	1965	1966	1967
Shipments (million tons)			
Pipelines	225.7	247.7	273.3
Crude oil	205.3	225.6	249.2
Products	20.4	22.1	24.1
Railroads	222.2	240.2	260.3
Crude oil only	n.a.	n.a.	n.a.
River	25.0	26.9	28.6
Sea	53.5	59.9	67.3
Coasting trade	26.1	26.0	28.3
Tankers in foreign trade	27.4	33.9	39.0
Turnover (billion ton-km)			
Pipelines	146.7	165.0	183.4
Railroads	280.4	301.9	326.7
Crude oil only	n.a.	n.a.	n.a.
River	28.6	30.5	30.9
Coasting trade	15.9	n.a.	17.2
Average length of haul (km)			
Pipelines	650	666	671
Railroads	1,262	1,257	1,255
River	1,114	1,134	1,080
Coasting trade	609	n.a.	608
Length of pipelines (thousand km, Jan. 1)	26.9	28.2	29.5
Crude oil	21.7	21.7	23.0
Products	5.2	6.5	6.5
By diameter (percent)			
Up to 350 mm	33.6	n.a.	30.3
351–500 mm	31.6	n.a.	37.5
501–700 mm	24.2	n.a.	22.1
701–800 mm	5.9	n.a.	5.8
801 and up	4.7	n.a.	4.3

n.a. = not available.

Sources: Pipeline shipments: 1966–71: *Transport i sviaz' SSSR* (*Transport and Communications of the USSR*), Moscow, 1972, p. 202; 1972–73: *Nar khoz;* 1975 Plan: *Transport i khranenie nefti i nefteproduktov*, 1972:12, p. 5.

Railroad shipments: 1966–72: *Nar khoz.*

Sea shipments: Data for sea shipments from *Nar khoz;* for coasting trade from *Transport i sviaz' SSSR* (*Transport and Communications of the USSR*), pp. 54, 141, 154–155, except 1968 from Ushakov, 1972, p. 9 and 1967 from *ENP,* 1970:12, p. 39. For a few years (1965, 1967) figures shown for coasting trade cover *malyi kabotazh* only (between ports within one ocean basin) but the amounts excluded (shipments between ports in different basins) is very small. Shipments by tankers in foreign trade is derived as the difference between all sea shipments and coasting trade.

Pipeline turnover: Nar khoz, except 1975 Plan from *Transport i khranenie nefti i nefteproduktov*, 1972:12, p. 5 and 1973 from *VS,* 1974:9, p. 91. We also know that of the total turnover on pipelines of 281.7 billion ton-km in 1970, 260 billion ton-km was crude oil, leaving 21.7 for products. The implied average length of haul (ALH) is 828 km for crude and 857 km for products (*PKh,* 1973:9, p. 60).

1968	1969	1970	1971	1972	1973	1975 Plan[a]
301.3	324.0	339.9	352.6	388.5	421.4	426
276.4	300.7	314.6	325.7	n.a.	n.a.	n.a.
24.9	23.3	25.3	26.9	n.a.	n.a.	n.a.
275.9	284.7	302.8	322.8	340.4	360.5	n.a.
n.a.	47.26	63.8	n.a.	n.a.	n.a.	n.a.
29.2	30.3	33.5	35.2	33.7	33.9	n.a.
70.1	70.5	75.1	79.8	83.1	84.2	n.a.
30.4	32.8	33.2	35.3	n.a.	n.a.	n.a.
39.7	37.7	41.9	44.5	n.a.	n.a.	n.a.
215.9	244.6	281.7	328.5	375.9	439.4	570
333.9	342.8	353.9	380.1	409.4	449.7	n.a.
n.a.	66.5	88.9	n.a.	n.a.	n.a.	n.a.
32.5	34.9	39.1	41.6	39.8	n.a.	n.a.
26.2	19.9	n.a.	n.a.	n.a.	n.a.	n.a.
717	755	829	932	968	1,042	1,338
1,210	1,204	1,169	1,178	1,203	1,252	n.a.
1,113	1,152	1,167	1,182	1,181	1,209	n.a.
862	606	n.a.	n.a.	n.a.	n.a.	n.a.
32.4	34.1	36.9	37.4	41.0	42.9	50
25.7	27.1	29.9	30.7	33.2	34.2	n.a.
6.7	7.0	7.0	6.7	7.8	8.7	n.a.
29.3	n.a.	25.8	25.2	n.a.	n.a.	n.a.
35.9	n.a.	26.4	27.4	n.a.	n.a.	n.a.
21.7	n.a.	27.2	26.9	n.a.	n.a.	n.a.
5.2	n.a.	7.9	7.7	n.a.	n.a.	n.a.
7.9	n.a.	12.7	12.8	n.a.	n.a.	n.a.

Railroad turnover: Nar khoz, except that the turnover for crude only is based on *PKh,* 1973:9, p. 60 for 1970, and on reported shipments and ALH for 1968, given in Gankin, *et al.* (eds.), 1972, p. 109.

River turnover: 1966–71: *Transport i sviaz' SSSR,* p. 166; 1972: by 1971 ALH and 1972 shipments.

Coasting trade turnover: 1965, 1967, 1969: *ENP,* 1970:12, p. 39 (these figures cover *malyi kabotazh* only); 1968: Ushakov, 1972, p. 9.

Average length of haul: Implied by shipments and turnover. 1969 is given in one source (*ENP,* 1970:12, p. 39) as 670 km; average length of haul is not shown separately for crude oil and products on pipelines as there is not much difference.

Length of pipelines: 1965–70: *ENP,* 1970:12, p. 40; 1971: *Transport i khranenie nefti i nefteproduktov,* 1972:12, p. 5; 1972–73: total from *Nar khoz,* and crude oil lines only from *EG,* 1973:5, p. 1; 1975 Plan: *Transport i khranenie nefti i nefteproduktov,* 1972:12, p. 5.

Size distribution of pipeline network: Rubinov, 1972, p. 28.

[a] Original Ninth Five-Year-Plan goals.

well beyond the original target: it will be about 59,800 kilometers on January 1, 1975, as compared with the 50,000 kilometers planned (*EG*, 1975:3).

The two questions most relevant to our contemporary concerns are the status and rate of change in pipeline technology and the cost of moving Western Siberian oil to the European USSR. Regarding pipeline technology, the indications are that there has been little innovation or productivity growth. Turnover on pipelines has increased at about the same rate as the length of the network, during a time when there has been a significant increase in the average diameter of oil pipelines. A very large share of additions to the oil pipeline network has been at the upper end of the distribution—28 inches and larger (see table 13). Similarly the cost per ton-kilometer for pipeline movement of oil reported in the statistical handbooks has stayed about level.

Although it was implied earlier that the Soviet Union is not likely to have much oil to export in the 1980s, since most of the Siberian oil will have to be used to cover domestic needs, it is still useful to get some ideas about what the cost of transportation from Western Siberia would do to the competitiveness of Siberian oil as an export, or as a fuel source in the European USSR.

The cost of pipeline transport of oil is subject to many variables. The Russians report an average cost per ton-kilometer on their oil pipeline network of about 0.1 kopek (*ENP,* 1970:12, pp. 38–41), but this is an average of the costs on many different kinds of lines, and does not include any charge for capital. For one of the biggest lines (unnamed, but accounting for about one-third of all pipeline oil shipments in 1969) the addition of an interest charge (at 10 percent) would have raised costs by 60 percent (*ENP*, 1970:9, p. 31). Because of the high capital intensity, depreciation charges account for about half the reported cost, and the addition of interest charges would thus make capital-related elements account for about 70 percent of cost. If a higher rate of interest is assumed, the share would be still higher. This is important, because the construction cost of pipelines varies significantly depending on terrain. In regions of high labor cost, or unfavorable terrain (rocky or marshy), construction cost can double (Ushakov, 1972, p. 143).

Perhaps the best perspective is given by the kind of planning calculations which the project-making organs have developed. One of the most thorough of the handbooks on the subject (Rubinov, 1972, p. 122) shows *privedennye zatraty* (a cost measure including an allowance for the opportunity cost of capital along with current operating costs) for oil transport as about .50 kopeks per ton-kilometer. This presupposes use of the largest diameter pipelines (1,220 mm or 48-inch), with no extra allowance for the costs expected for northern lines. For the pipeline from Samotlor to Al'metevsk (about 2,100 km long, which is still not as far west as the Volga), this would be about 10 rubles per ton. Remember that, in table 12, extraction cost in Western Siberia was shown as about 3.40 rubles per ton in 1971 (without rent or capital charge), so that even if these costs are corrected to include an allowance for capital, transportation more than doubles the cost in the European USSR. This is probably still a relative bargain in energy transport, however, compared with the costs of moving Siberian gas, coal, or hydroelectric power to the European USSR.

6

REFINING AND UTILIZATION OF PETROLEUM PRODUCTS

Most of the distinguishing features of Soviet petroleum refining and of Soviet petroleum product consumption described in *ESOG* remain unchanged. The USSR has a relatively simple refinery mix, which reflects a relatively undemanding market and technological backwardness in refining. A large share of total consumption of petroleum products is in the form of residual fuel oil rather than light products, and the general quality level of products remains low. The highest priority of the industry continues to be quantitative expansion to handle the rising volume of oil rather than to intensify refining and upgrade quality.

Refining is a difficult sector of the Soviet oil industry to study because of the lack of data on refinery operations and on consumption. There has been even less information released for recent years than there was for earlier years.

In table 14, the amounts of oil refined are estimated on the basis of crude output, corrected for field losses, exports, and filling of new pipelines. Further allowance for fuel use and losses in refining gives an estimate of total refinery products output, and correction for trade in petroleum products then shows the total consumption of petroleum products. There is a great deal of scattered evidence, much of it unclear and internally contradictory, regarding the composition of consumption and production by product, and increases in output of various products. This evidence is used to reconstruct consumption and production of the main products for various years in table 15. Because the evidence is sparse and inconsistent these are tentative estimates that could be appreciably in error, but they provide a reasonably sound basis for characterizing the nature of Soviet consumption and for reviewing certain trends. A comparison of Soviet, U.S., and Western European petroleum product consumption appears below.[1]

[1] Soviet figures from VNIGRI, 1973a, p. 55; U.S. figures from U.S. Bureau of Mines, *Minerals Yearbook, 1971,* GPO, Washington, D.C., 1972, pp. 878–879; Western European figures from OECD, *Oil Statistics, Supply and Disposal, 1972,* Paris, 1973, p. 14.

Product	USSR 1975	U.S. 1970	Western Europe 1970
	(.	Percent of total	. . . .)
Gasoline	19.9	39.7	14.9
Kerosene and jet fuel	8.6	8.4	6.0
Diesel fuel	24.7	17.3	31.2
Residual fuel oil	36.1	15.0	39.6
Other	10.7	19.6	8.3

What is striking is the relative similarity between USSR and Western European consumption. Indeed, Western European gasoline consumption is even lower, and residual fuel oil consumption even higher, than in the USSR.

The high share of residual fuel is attributable to Soviet efforts to reap the cost advantage of petroleum as a fuel for boiler and furnace use, and to the inability of the industry to increase the depth of refining appreciably. The low share of gasoline in Soviet consumption relative to the U.S. pattern is attributable to the relative unimportance thus far of the passenger automobile. The relatively high share of kerosene is a result of the anachronistic combination of jet fuel demand with sizable consumption of kerosene in tractors and households. The share of kerosene has been falling fairly rapidly, however, since 1965.

Industry planners affirm a desire to raise the share of light products, that is, to get more diesel fuel and gasoline out of a barrel of oil by more secondary processing. This aim has foundered on continued failure to attain planned growth in secondary refining capacity. In 1960–65 the plan for capacity increases in various kinds of secondary processing facilities was fulfilled by only 24 to 33 percent (*EG*, 1966:17), and in 1965–70 also, V. S. Fedorov, Minister of the Petroleum Refining and Petrochemicals Industry, said that the industry managed to cope with growth in the amount of oil to be refined only by letting the share of residual fuel oil rise above the share planned (*Neftepererabotka i neftekhimiia,* 1970:4, p. 203).

Quality Improvements in Gasoline and Diesel Fuel

The most significant improvement in the refining branch has been in the quality of individual products, notably gasoline and diesel fuel, as shown in table 16.

The extent of improvement in gasoline quality should not be exaggerated. The goal has been to shift from an output consisting predominantly of grade A-66 to one that is mostly grades A-72 and A-76.[2]

[2] In these grade designations A stands for automotive, the number for motor-method octane ratings. Grade A-72 differs from A-66 also by a somewhat lighter

TABLE 14. ESTIMATE OF SOVIET CONSUMPTION OF PETROLEUM PRODUCTS
(million tons)

Item	1965	1966	1967
Production of crude and condensate	242.9	265.1	288.1
Less exports of crude	43.4	50.3	54.1
Plus imports of crude	neg.	neg.	neg.
Less 5% of output (field loss)	12.1	13.3	14.4
Less filling new pipelines[a]	0.2	0.2	0.5
Refinery runs	187.1	201.3	219.1
My index (1965 = 100)	100	107.6	117.1
Official Soviet index	100	106.7	116.0
Products output[b]	164.6	177.2	192.8
Less products exports	21.0	23.3	24.9
Plus products imports	1.9	1.7	1.4
Apparent consumption	145.5	155.5	169.3

n.a. = not available. neg. = negligible.

Sources: Production: Table 9.

Field losses: Field losses of 5 percent as in *ESOG*. Some evidence found after these calculations were finished suggests that losses may be less—about 3 percent. Fedorov says that on April 5, 1970, the petroleum refining industry refined the billionth ton of oil during the Ninth Five Year Plan. If losses are estimated as 5 percent, the amount that would have to be refined from Jan. 1 to April 5, 1970, to reach 1 billion tons is so large that the implied output for 1970 is impossibly large; it is more oil than is available. The 3 percent loss rate, however, would generate an amount for this period that would imply a reasonable total for 1970.

Losses and fuel costs: There are many partial indications of refinery fuel and losses. I have chosen 9 percent for fuel on the basis of what seems to be an elaborate and comprehensive analysis in *Khim,* 1974:3, pp. 28–30, that implies 9.8 percent. Some of this fuel expenditure involves fuels other than oil, and so I adjust it

But even when 66 octane gasoline has been eliminated entirely and the bulk of gasoline is 72 octane, this will still represent a rather low quality standard. It is far below the general level of octane ratings in the United States (around 90, motor method, for *regular* grade gasoline), and can hardly be an optimal quality level. Soviet economists say that cost savings from raising the average octane rating beyond 72 to make possible fuel and performance improvements through higher compression ratios would far exceed the incremental refinery cost (Umanskii and Umanskii, 1974, p. 134). The explanation for the failure to do so seems to be capital shortages and technological weakness.

fractional composition, less gum, and higher stability. It is also nonleaded. Grade A-76, except for the slightly higher octane rating, is basically like A-72, but contains lead (*Tovarnye nefteprodukty,* 1971). Given that the average octane rating for all gasoline output in 1970 was reported as 69 (which could be represented by a half-and-half mixture of A-66 and A-72) the share of grades above 72 octane must be balanced by continued output of the older grades even below A-66, such as A-56.

1968	1969	1970	1971	1972	1973	1974
309.2	328.4	353.0	377.1	400.4	428.6	458.8
59.2	63.9	66.8	74.8	76.2	85.3	80.6
0.1	1.4	3.5	5.1	7.8	13.2	4.4
15.5	16.4	17.7	18.9	20.0	21.4	22.9
0.3	0.6	0.1	0.8	0.4	0.9	2.7
234.3	248.9	272.0	287.8	311.6	334.1	357.2
125.2	133.0	145.4	153.8	166.5	178.6	190.1
123.5	132.0	144.0	152.4	163	n.a.	n.a.
206.2	219.0	239.3	253.2	274.2	294.0	314.3
27.0	26.9	29.0	30.3	30.8	33.0	35.6
1.1	1.1	1.1	1.5	1.3	1.5	1.0
180.3	193.3	211.4	224.4	244.7	262.5	279.7

down to 9 percent. The 3 percent losses figure is based on a statement in *Trud,* Sept. 13, 1968.

The Soviet index for refinery runs is based on statements that: 1970/1965 = 1.44 (Umanskii and Umanskii, 1974, p. 29); 1966/1965 = 1.067 (*Khim,* 1966:11); 1967/1965 = 1.16 and 1968/1972 = 1.065 (*Neftepererabotka i neftekhimiia,* 1969:1); 1969/1965 = 1.32 (*Khim,* 1970:1); 1971 Plan/1970 = 1.058 (*Khim,* 1971:1); and 1972 Plan/1971 = 1.07 (*Khim,* 1972:4).

[a] Figured on basis of average d = 20″ in 1965–68; average d = 23″ in 1969–73. This leads to a slight change in amount to fill new pipelines and in the results of all subsequent steps, compared to the 1965 calculation shown in *ESOG.* In some cases components may not add to totals because of rounding.

[b] Calculated as refinery runs less 9 percent for fuel use and less 3 percent for refinery losses.

There is a hint here that this might be an area in which the USSR will seek technological transfer from the advanced countries.

The reduction in sulfur content of diesel fuel is more impressive. As explained in *ESOG,* in the early sixties the standards specified variously 0.2 percent, 0.6 percent, and 1 percent sulfur content, but in fact the majority of all diesel fuel was produced with a sulfur content of 1 percent or above. This meant poor performance, high corrosion, and expensive maintenance. The strategy chosen to deal with the problem seems to have been to set, in 1963, a new standard of 0.5 percent sulfur content, presumably low enough to permit controlling corrosion with lube additives. Progress toward getting all diesel fuel within that goal was fairly quick—from 39.3 percent of the total in 1965 to over 90.5 percent in 1972. It was promised that by 1975 virtually all diesel fuel would meet the 0.5 percent standard.

Both the increase in octane rating and the reduction in sulfur content have been greatly enhanced by a change in the composition of crudes to be refined. Western Siberian crude, which has been rising rapidly in

TABLE 15. SOVIET CONSUMPTION AND PRODUCTION OF PETROLEUM PRODUCTS, BY PRODUCT

(million tons)

	Consumption	Production			
	1970	1965	1969	1970	1973
Gasoline	41.2	31.0	39.6	43.0	58.6
Auto gas	n.a.	29.4	n.a.	n.a.	n.a.
Kerosene	20.5	15.0	n.a.	22.0	24.7
Diesel fuel	50.8	44.0	57.0	62.0	78.9
Subtotal	112.5	90.0	n.a.	127.0	n.a.
Other light products	4.7	4.0	n.a.	4.5	4.8
Total light products	117.2	94.0	n.a.	131.5	167.0
Lube oils	5.3	4.0	[a]	5.6	[a]
Residual fuel oil	74.0	60.0	n.a.	87.2	105.2
Furnace grade	66.6	54.0	69.3	78.3	n.a.
Other	7.4	6.0	n.a.	8.9	n.a.
Other heavy products	14.9	6.6	n.a.	15.0	21.8
Total heavy products	94.3	70.6	n.a.	107.8	127.0
Total products	211.5	164.6	n.a.	239.3	294.0

n.a. = not available.

Sources: Consumption in 1970 is derived by applying to total consumption in 1970 (as estimated in table 14) a percentage structure derived from VNIGRI, 1973a, p. 55, and from the structure of consumption in 1958 as shown in *ESOG,* table 24. Correction for trade in the various products (available in the Soviet foreign trade handbooks), converts 1970 consumption into 1970 production, but what is shown in the table is an adjustment of that result to make it more consistent with other evidence. Indexes of production increases for the various categories of products from various issues of *Neftepererabotka i neftekhimiia* and *Khim* are then used to estimate production in other years.

A detailed description of the evidence, the estimating procedure, and the rationale for resolving contradictions among sources is too complex to justify exposition here, but is available on request to the author.

The production figures for 1965 given in this table differ from those for the same year in *ESOG,* table 25, but are preferable because they are more consistent with the evidence on production and consumption for 1965–73 used in constructing this table.

[a] Included in other heavy products.

relation to that from the Volga–Ural fields, has a lower sulfur content and better gasoline properties than the Volga–Ural crudes. The other new sources of output—Komi ASSR, Belorussia, and Krasnodar Krai—also produce crudes of higher quality. The quality of the oil produced from the Mangyshlak fields in western Kazakhstan, the other new element in the total, is less favorable.

PROSPECTIVE CHANGES IN REFINERY MIX

There is a question whether continuation of this distinctive pattern of refining and consumption in the USSR is a rational use of Soviet crude

TABLE 16. QUALITY IMPROVEMENT IN GASOLINE AND DIESEL FUEL
(percent)

Year	Share of automotive gasoline with octane rating 72 and higher	Share of diesel fuel with sulfur content 0.5 percent and less
1965	14.8	39.3
1966	20.7	48.2
1967	20.0	n.a.
1968	n.a.	70.0
1969	40.0	73.0
1970	50.0[a]	80.0
1971	69.0	n.a.
1972	70.5	95.5

n.a. = not available.

Sources: Khim, 1969:1; 1970:11; 1971:1; 1971:10; 1973:12; *Standarty i kachestvo,* 1968:4, p. 35, *Sotsialisticheskaia Industriia,* August 19, 1973.

[a] Other evidence (statements by the Minister of the Petroleum Refining and Petrochemical Industry in *Neftepererabotka i neftekhimiia,* 1970:4, and 1971:4) suggests that this 50 percent figure is exaggerated.

oil resources, given the changing world prices for energy products. It seems unlikely that it is, and it will be interesting to see how rapidly the Soviet planners move to adjust it.

First we can expect the growing importance of the private automobile in the USSR to have a serious impact on the pattern. The Soviet automotive stock is now growing rapidly. Passenger automobile output was about 250,000 per year in the sixties, of which about 150,000 were added to the domestic stock. By 1975, output was to be about 1.25 million units per year, and even with significant exports the size of the increments in relation to the stock will be far above that of the recent past.

The slow upward creep of the share of residual fuel oil in products consumption that has taken place since the fifties means that crude oil output and refining were being expanded faster than necessary to meet the need for light products only. There is an earlier Soviet idea, dominant in the thirties and forties, that using oil for boiler fuel is wasteful. This idea was abandoned in the new fuel policy adopted in the late fifties, but Soviet energy policy seems once again to be returning to the older idea. An old quotation from Mendeleev, that burning oil in boilers is equivalent to burning banknotes, has been resurrected and was frequently cited during 1974. An increase in the depth of refining to meet expansion of demand for products not replaceable by other fuels would reduce the pressure on expansion of crude oil output, or release more crude oil for export. But if it is decided to continue petroleum exports it would be better to do so in the form of products, rather than crude oil, to obtain higher foreign exchange earnings. One would judge from the present mix that export potential for products is much constrained by quality. The biggest item in the current mix is diesel fuel. Gasoline and kerosene have a small share, and residual fuel oil has usually made up about 40 percent of total products exports.

The biggest domestic consumer of residual fuel oil has been the electric power industry, and its ability to shift to other fuels may be very limited in the short run. Electric power plants have long gestation periods, and most of those now being built in the European USSR are designed to burn residual fuel oil. It will take a long time to get much relief through nuclear power.

It is also interesting to see how insignificant are petrochemical feedstocks in the Soviet product mix. Petrochemical feedstocks are supplied by both gas processing plants and petroleum refineries. In petroleum refineries alone, petrochemical feedstocks constitute about 10 percent of output in the United States but constitute a negligible share of such output in the USSR. The one distribution I have seen (that projected for 1975 and used above in deriving the Soviet refinery output mix) does not distinguish petrochemical feedstocks as a separate category, except for aromatics (0.8 percent of the total) and carbon-black feedstocks (0.4 percent of the total). Significant amounts of low octane gasoline or kerosene may be used for petrochemical raw material. Soviet petroleum refineries apparently produce only small amounts of liquefied refinery gases, including ethane and ethylene, because they use their still gas mostly for refinery fuel and because, with the relatively slight emphasis on secondary processing, refinery gases are not produced as abundantly as in American refineries. It is also said that such secondary processing as is done occurs with a technology that minimizes the output of products usable as petrochemical feedstocks.

To alter and improve its product mix, the refining branch of the industry will require large capital expenditures and an accelerated effort to improve technology, perhaps by technological transfer from abroad.

7

GAS INDUSTRY

The Soviet Union is blessed with very rich natural gas resources, and has increased gas output to a point where it has become an important element in the fuel balance. Although gas reserves are large enough to be a serious potential source for energy exports, the gas industry has consistently fallen behind planned growth, and, because of the location of its resources, faces serious technological problems in production and transport.

RESERVES

Soviet natural gas reserves are adequate for maintenance and expansion of output for many years to come. Since 1965 the Russians have proved up large amounts of natural gas to the $A+B+C_1$ standard. As shown in the following tabulation of reserves in these categories (in trillion cubic meters as of January 1 of each year) they rose from 3.6 trillion cubic meters at the beginning of the Eighth Five Year Plan period (January 1, 1966) to 24.6 trillion cubic meters at the beginning of 1975.[1] In comparison, U.S. proved reserves of natural gas were

1966	3.6	1971	15.8
1967	4.4	1972	18.0
1968	7.8	1973	19.5
1969	9.5	1974	22.4
1970	12.1	1975	24.6

estimated to be a little over 7 trillion cubic meters in 1974 (Federal Energy Administration, 1974, p. 89).

In addition to $A+B+C_1$ reserves, it is estimated that as of January 1, 1971, there were 68 trillion cubic meters of reserves in the more speculative C_2, D_1, and D_2 categories (*GP,* 1970:1). These are probably very conservative forecasts, since reserves in these categories have been

[1] From the following sources: 1966–67: Kortunov, 1967, pp. 33, 34, 55; 1968: L'vov, 1969, p. 84; 1969: *GP,* 1970:1, p. 4; 1970: *GP,* 1970:4, p. 10; 1971, 1973: Brents, 1975, p. 25; 1972: *GP,* 1972:2, p. 1; 1974: *GP,* 1975:1, p. 5; 1975: *GP,* 1975:6, p. 2. The figure for 1966 differs from that for the same date cited in *ESOG,* and is from a later source.

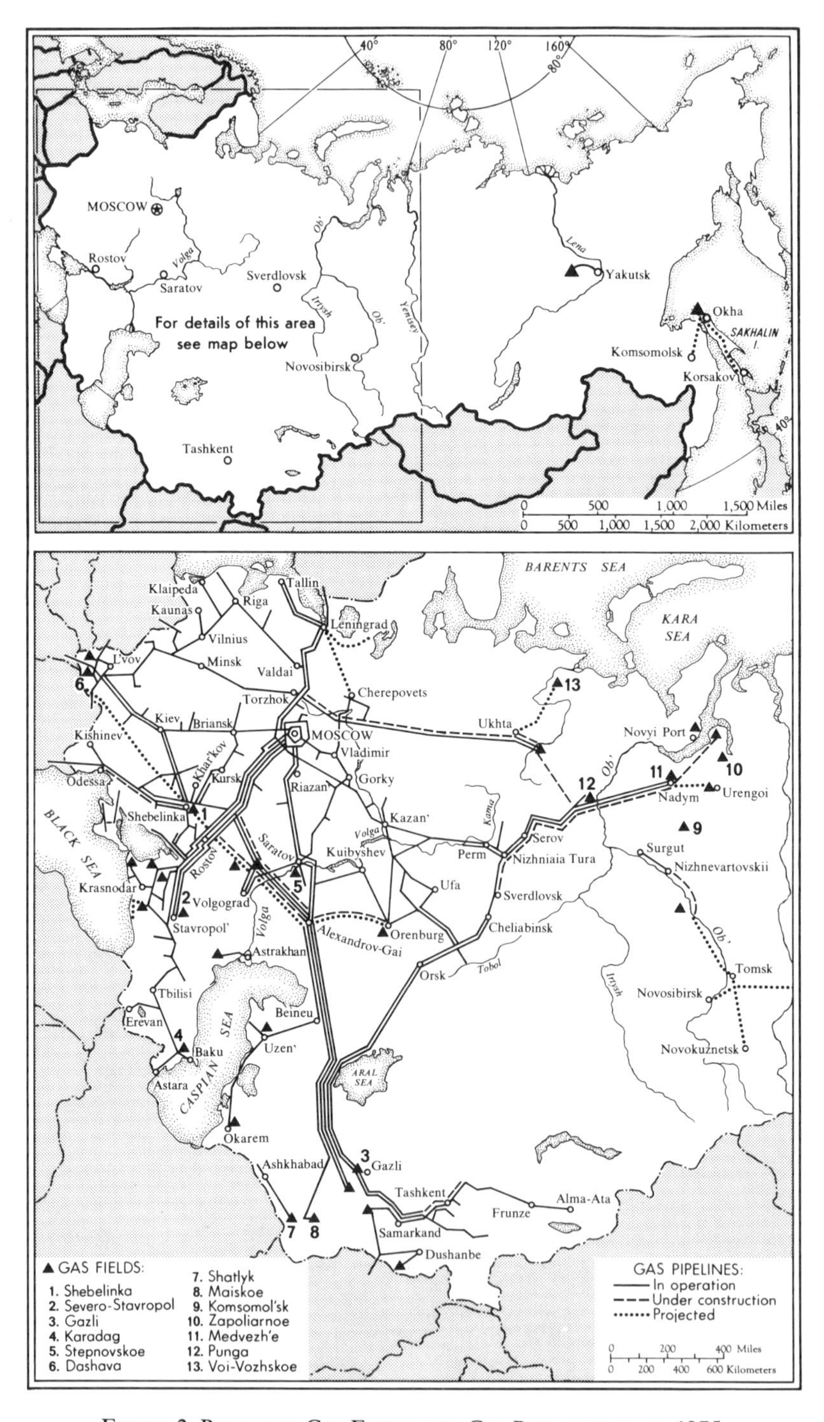

FIGURE 2. PRINCIPAL GAS FIELDS AND GAS PIPELINES AS OF 1975

estimated for only 58 percent of the promising territory (L'vov, 1969, p. 75). Nor do they include any allowance for gas reserves in shelf areas, which one source estimates at 22.7 trillion cubic meters (*GP*, 1970:1, p. 5).

These impressive gains in $A+B+C_1$ reserves were won at very low cost in exploratory activity. About 10 million meters of exploratory drilling for natural gas in the Eighth Five Year Plan period raised reserves by about 13 trillion cubic meters. The reserves are contained mostly in a small number of very large fields, and this large average size of new discoveries explains the relatively low cost and rapid growth of reserves.

Unfortunately, the most significant additions to Soviet reserves of natural gas have been in disadvantageous locations, especially Western Siberia and Central Asia.[2] As table 17 shows, the USSR's gas riches are located overwhelmingly in Tiumen' oblast' in Western Siberia. More recent reports show an even more lopsided distribution. On January 1, 1974, 63 percent of all $A+B+C_1$ reserves were in Tiumen' oblast', another 15.4 percent in other Asian regions of the USSR, mostly Central Asia and Kazakhstan, and only 16 percent in the European USSR (*GP*, 1975:1, p. 5). Older areas in the European USSR, which were the sources of the rapid expansion of the gas industry during the sixties, have experienced very little increase. Indeed, some areas in the Volga–Ural and North Caucasus regions have suffered declines in reserves as their big fields were drawn on heavily for production, while no new finds were made. It is unlikely that this location problem will be much eased by future exploration—the *prognoznye* reserves (categories D_1 and D_2, those considered probable, but not yet found) show essentially the same locational pattern as those already explored, as shown in the following comparison of the regional distribution for 1970 (*GP*, 1970:1, p. 4).

	$A+B+C_1$	D_1+D_2
	(percent of total)	
European areas	22	26
Western Siberia	59	29
Eastern Siberia and Far East	3	28
Central Asia and Kazakh SSR	16	17

The main difference is that the probable reserves are distributed in a pattern skewed even more eastward than the explored reserves.

The major exception to the generalization above is the fairly recent discovery of large fields in the Komi ASSR and in Orenburg oblast'. The Vuktyl field in the Komi ASSR is thought to have reserves of 500 billion cubic meters, and the Orenburg field over 2 trillion cubic meters.

[2] For regional references in this chapter see the maps opposite p. 1 and on p. 50.

TABLE 17. GROWTH AND GEOGRAPHIC DISTRIBUTION OF SOVIET NATURAL GAS RESERVES

(billion cubic meters)

Region	1966	1970	Increment, 1966–70
USSR total	3,566	12,092	8,526
RSFSR, European	1,093	1,846	753
Komi ASSR	38	383	345
Bashkir ASSR	30	54	24
Perm oblast'	24	40	16
Kuibyshev oblast'	11	6	−5
Orenburg oblast'	25	537	512
Saratov oblast'	71	73	2
Volgograd oblast'	90	88	−2
Krasnodar krai	465	365	−100
Stavropol' krai	234	223	−11
Chechen-Ingush ASSR	9	8	−1
Dagestan ASSR	43	46	3
Other	53	23	−30
RSFSR, Asian	602	7,450	6,848
Sakhalin oblast'	49	72	23
Tiumen' oblast'	401	6,774	6,373
Tomsk oblast'	54	234	180
Iakutsk ASSR	78	277	199
Other	20	93	73
Ukrainian SSR	655	726	71
Azerbaidzhan SSR	54	54	0
Kazahk SSR	92	183	91
Uzbek SSR	667	730	63
Turkmen SSR	376	1,053	677
Tadzhik SSR	13	32	19
Kirgiz SSR	14	17	3
Georgian SSR	0	1	1

Note: These are $A+B+C_1$ reserves.
Sources: 1966: Kortunov, 1967, p. 36; 1970: *GP*, 1970:4, p. 10.

Performance on output growth since 1965 has been less encouraging than on reserves. As shown in table 18, output has grown at fairly high rates—an average of 8.2 percent per year in the period 1965–74. Gas now accounts for about 20 percent of all primary energy production, and about 28 percent of fuel used in boilers and furnaces. Nevertheless, output growth has been below the much higher rates in the decade centering on 1960, and much below plan. For example, the original target of 320 billion cubic meters for 1975 has now been adjusted downward to 285 billion cubic meters.

To exploit its gas riches, the USSR must shift the regional production pattern from that of the 1960s, which drew very heavily on the fields in the European part of the USSR, to one that brings gas to market in the European USSR over long distances from Siberia and Central Asia. As

TABLE 18. GAS OUTPUT

(billion cubic meters)

Year	Natural gas output utilized			Oil-well gas wasted
	Gas-well	Oil-well	Total	
1965	111.2	16.5	127.7	7.1
1966	125.2	17.8	143.0	7.8
1967	138.6	15.9	154.4	9.0
1968	149.5	19.6	169.1	10.9
1969	159.5	21.6	181.1	12.4
1970	174.9	23.0	197.9	14.6
1971	187.4	25.0	212.4	15.9
1972	195.6	25.8	221.4	n.a.
1973	209.8	26.5	236.3	n.a.
1974	233.0	27.6	260.6	19.0
1975 (Plan)	n.a.	n.a.	285.2	n.a.
1975 (9th FYP)	n.a.	n.a.	320.0	n.a.

n.a. = not available.

Sources: Gas-well gas: 1965–70: *GP,* 1971:5; 1971, by subtraction; 1972–73: Brents, *et al.,* 1975, p. 39; 1974, by subtraction.

Oil-well gas: 1965–70: total, minus gas-well gas; 1971: *Gazovoe delo,* 1972:4, p. 32; 1972–73: by subtraction; 1974: Brents, *et al.,* 1975, p. 33.

Total: 1965–73: *Nar khoz;* 1974: Brents, *et al.,* 1975, p. 33; 1975 Plan: *GP,* 1975:1, p. 2; 1975 9th Plan: Baibakov, 1972.

Oil-well gas wasted: 1965: *ENP,* 1973:1, p. 4; 1966: Luzin, 1974, p. 115; 1967: *Gazovoe delo,* 1968:8, p. 24; 1968–70: *Gazovoe delo,* 1972:4, p. 33; 1971: *GP,* 1971:5; 1973: *EG,* 1975:3, p. 2.

table 19 shows, this transition is already well under way, though the present geographical distribution of output is quite different from that of reserves.

For the moment the strategy chosen for minimizing the difficulties in this transition is to concentrate on several of the largest of the new fields that pose the fewest problems and to postpone until later the exploitation of the bigger but more difficult ones in northern Tiumen' oblast'. Through 1975 output increases were based primarily on Medvezh'e in Western Siberia, Vuktyl in Komi ASSR, Shatlyk and Naip in the Turkmen SSR and Orenburg in Orenburg oblast'. These fields, which produced virtually no output in 1970, were scheduled to provide 132 billion cubic meters of the planned 1975 output of 320 billion cubic meters (*GP*, 1972:2). But even these fields present problems of production and transport technology. Orenburg is well located but it is a high sulfur and condensate field, a situation with which the Soviet gas industry has not had a lot of experience. Shatlyk is distant, though it can be hooked up to the already existing Central Asia–Moscow pipeline complex. Medvezh'e is in the north and raises problems of production and transport in the permafrost area.

TABLE 19. REGIONAL DISTRIBUTION OF NATURAL GAS OUTPUT
(billion cubic meters)

Region	1965	1966	1967	1968
Total, including oil-well gas	127.7	143.0	157.4	169.1
RSFSR	64.3	69.0	74.8	78.3
Komi ASSR	0.8	0.9	0.9	0.9
Volga–Ural	22.8	23.4	n.a.	n.a.
West Siberia	a	a	n.a.	n.a.
North Caucasus	39.9	44.1	n.a.	n.a.
Far East	0.6	0.6	n.a.	n.a.
Ukrainian SSR	39.4	43.6	47.4	50.9
Azerbaidzhan SSR	6.2	6.2	5.8	5.0
Uzbek SSR	16.5	22.6	26.6	29.0
Kirgiz SSR	0.1	0.2	0.3	0.3
Tadzhik SSR	a	0.1	0.2	0.4
Turkmen SSR	1.2	1.3	2.2	4.8
Kazakh SSR	a	a	0.1	0.3
Belorussian SSR	0.0	0.0	a	a
Total, excluding oil-well gas	111.2	125.2	138.6	149.5
RSFSR	53.7	57.3	62.0	64.6
Komi ASSR	0.6	0.6	0.6	0.6
Volga–Ural	15.3	14.8	13.3	10.6
West Siberia	a	0.6	0.2	8.2
North Caucasus	37.5	41.0	42.6	44.7
Far East	0.3	0.3	0.4	0.5
East Siberia	0.0	0.0	0.0	0.0
Ukrainian SSR	37.8	41.9	45.6	49.0
Azerbaidzhan SSR	3.0	3.1	2.9	2.5
Uzbek SSR	16.4	22.5	26.5	28.9
Kirgiz SSR	0.1	0.2	0.2	0.3
Tadzhik SSR	a	0.1	0.2	0.4
Turkmen SSR	0.1	0.2	1.1	3.6
Kazakh SSR	a	a	a	0.2

Notes: n.a. = not available. In some cases components do not add to totals because of rounding. Regional detail within the RSFSR is often inconsistent with the RSFSR total, due to inconsistencies in the sources. Information for reconciling the two is not available.

Sources: Union-republic breakdown: Data for total gas output are given in *Nar khoz* through 1973. Additional information in some republican handbooks, and in *GP*, 1970:4 and 1971:5 permits a division into gas-well (*prirodnyi*) gas and oil-well (*poputnyi*) gas. For the years after 1970, information is more sketchy, and the numbers in the table are from a variety of sources—republic handbooks; Lisichkin (ed.), 1974a, p. 68, and Brents, 1975, p. 39. Data for 1975 Plan are from Baibakov, 1972.

The contrast between the abundance of reserves and the failure to meet targets is a recurrent theme in Soviet analyses of gas industry performance. As one Soviet official says, "as a practical matter, the development of the gas extraction industry in the near term is not constrained by its raw material base, but is determined by its technical–economic possibilities" (*GP*, 1975:5, p. 6).

1969	1970	1971	1972	1973	1975 Plan
181.1	197.9	212.4	221.4	236.3	320.0
81.0	83.3	87.5	87.4	87.8	144.6
2.2	6.9	10.5	n.a.	n.a.	16.1
n.a.	18.6	n.a.	n.a.	n.a.	n.a.
n.a.	9.3	11.0	n.a.	n.a.	44.0
n.a.	47.1	n.a.	n.a.	n.a.	16.0
n.a.	1.0	n.a.	n.a.	n.a.	n.a.
55.4	60.9	64.7	67.2	68.2	62.0
4.9	5.5	5.8	6.9	8.4	6.0
30.8	32.1	33.7	33.7	37.1	33.7
0.3	0.4	0.4	0.4	0.4	[a]
0.4	0.4	0.4	0.5	0.5	0.5
7.5	13.1	16.9	21.3	28.6	65.1
0.7	2.1	2.7	3.5	4.8	6.2
[a]	0.2	0.3	0.4	0.4	0.6
159.5	174.9	187.4	195.6	209.8	n.a.
65.7	66.8	69.9	69.5	69.9	n.a.
1.9	6.5	10.0	13.0	15.9	n.a.
10.0	8.9	10.5	10.2	8.4	n.a.
9.1	9.2	9.3	11.4	15.8	n.a.
44.1	41.0	38.3	32.0	26.7	n.a.
0.6	0.8	0.8	0.8	0.9	n.a.
0.0	0.4	1.0	2.2	2.4	n.a.
53.1	58.6	62.2	64.7	65.8	n.a.
2.4	3.0	2.9	3.7	4.8	n.a.
30.7	32.0	33.5	33.6	37.0	n.a.
0.3	0.4	0.4	0.4	0.4	n.a.
0.4	0.4	0.4	0.5	0.5	n.a.
6.3	11.8	15.5	19.8	27.1	47.5
0.6	2.0	2.6	3.4	4.3	n.a.

Regional breakdown within RSFSR, total: There is some information in the RSFSR handbooks, and in the regional handbooks. The breakdown for 1970 is based on Baibakov, 1972, and Umanskii and Umanskii, 1974, p. 171; 1968, Komi: Parashchenko, 1972, p. 43; 1971, West Siberia: N. A. Eremenko, *et al. AAPG Bulletin,* September 1972, p. 1715; 1975 Plan from Baibakov, 1972, and *GP,* 1975:6, p. 2.

Regional breakdown within RSFSR, excluding oil-well gas: 1965, 1966, and 1967: L'vov, 1969, pp. 26, 27; 1968–71: Lisichkin (ed.), 1974a, p. 68, Brents, 1975, p. 39, and *Gazovoe delo,* 1972:4, p. 32; 1972 and 1973: Brents, 1975, p. 39.

[a] Less than 100 million cubic meters.

The "technical–economic" constraint the author has in mind is essentially that of mastering the huge investment programs and developing the technology to deal with new situations that production and long distance transport from the Siberian fields involve.

In remote and isolated areas the absence of transport and other infrastructural facilities and the presence of permafrost greatly complicate the job of developing production. For pipelines, problems are posed not only by remote and inhospitable terrain, but also by the need to develop very large-capacity lines. To handle the volumes of gas involved, there is a big advantage in developing pipelines of capacities not yet mastered, using pipe diameters of 48 inches and above. For such pipelines, the pipe, construction equipment, and improved control equipment for compressors must all be newly developed. In short, the Soviet gas industry is working at the frontier of technology in inhospitable circumstances. Significant amounts of gas are now being produced from some of these difficult areas, and considerable experience has been gained in building and operating pipelines from the Arctic areas. But the Soviet gas industry is a long way from really handling the problems successfully, and is still very much interested in getting assistance from the advanced capitalist countries in forms ranging from import of finished items such as large diameter pipe to more intimate forms of cooperation such as the proposed project for joint U.S.–USSR

TABLE 20. SELECTED INDICATORS OF GAS PIPELINES AND STORAGE

Indicators	1965	1966	1967
Length of pipeline in operation (1,000 km)			
Transmission lines	41.8	47.4	52.6
Gathering lines	4.4	4.4	5.7
Distribution lines	34.5	39.1	43.8
Installed compressor capacity (1,000 hp)	2,504	3,264	3,296
By prime mover (percent)			
Gas turbines	40	37	n.a.
Electric	48	46	n.a.
Piston	12	17	n.a.
Gas deliveries (billion m^3)	112.6	128.8	143.3
Average length of haul (km)	680	701	781
Storage (billion m^3)			
Active capacity	n.a.	n.a.	2.2
Injected	1.8	2.2	3.2
Withdrawn	1.0	1.4	1.8

Notes: n.a. = not available. Figures for stock concepts refer to December 31 of year shown.

Sources: Transmission lines: Nar khoz, various years, except 1975 Plan from Baibakov, 1972.

Gathering lines: 1965–66: Gal'perin, 1968, p. 40; 1967–71: *Gazovoe delo,* 1972:4, p. 35.

Distribution lines: 1965–66: Gal'perin, 1968, p. 40; 1967–71: *Gazovoe delo,* 1972:4, p. 35.

Installed compressor capacity: 1965–68: Gal'perin, 1968, p. 34; 1970: *GP,* 1971:1, p. 1; 1974: *GP,* 1975:5, p. 8; 1975 Plan: *GP,* 1975:6, p. 2.

Distribution by prime mover: 1965–66: Gal'perin, 1968, p. 36; 1967: Kortunov, 1967, p. 114; 1971, 1975 Plan: *Stroitel'stvo truboprovodov,* 1971:3, p. 22; 1974: *GP,* 1975:5, p. 8.

development of northern Tiumen' oblast' gas for shipment in liquefied form to the United States.

Gas Pipelines

The main indicators of the development of Soviet gas pipelines are shown in table 20. The network has doubled in length since 1965, with average annual additions of more than 5,000 kilometers. The Soviet pipeline network has always been characterized by an emphasis on large-diameter pipe, and this emphasis has been strengthened in recent years. On January 1, 1964, the largest pipe used was 40-inch, which accounted for 11.2 percent of the total. In 1969 the first 48-inch line was commissioned, and by 1972 there were 4,422 kilometers of 48-inch line (about 6.1 percent of the total) in operation. The heavy emphasis on large diameter line is shown by a comparison with the U.S. network (see table 21). The contrast shows up most strikingly in the fact that almost one-fourth of the Soviet total was 40- and 48-inch line, while in the United

1968	1969	1970	1971	1972	1973	1974	1975 Plan
56.1	63.2	67.5	72.5	79.0	83.5	n.a.	100.0
7.2	8.3	11.3	14.2	n.a.	n.a.	n.a.	n.a.
49.7	55.8	62.6	68.8	n.a.	n.a.	n.a.	n.a.
3,381	n.a.	4,020	n.a.	n.a.	n.a.	8,847	11,300
n.a.	n.a.	52	n.a.	n.a.	n.a.	67	68
n.a.	n.a.	29	n.a.	n.a.	n.a.	n.a.	16
n.a.	n.a.	19	n.a.	n.a.	n.a.	n.a.	16
155.1	166.0	181.5	209.8	219.9	231.1	n.a.	n.a.
935	970	917	n.a.	n.a.	n.a.	1,154	n.a.
n.a.	6.0	6.3	n.a.	7.0	n.a.	n.a.	n.a.
3.8	4.1	5.0	5.5	n.a.	8.1	10.4	13.0
2.7	3.4	3.6	3.7	n.a.	n.a.	n.a.	n.a.

Gas delivered: Nar khoz, various years. Another series, narrower in coverage, was reported earlier—it apparently included only deliveries via the transmission lines of the Ministry of the Gas Industry whereas this one includes other lines as well. The difference in 1968 was about 10 billion m^3.

Average length of haul: Zhigalova, 1973, p. 92, except 1970: *GP,* 1975:1, p. 5; and 1974: *GP,* 1975:5, p. 8.

Storage capacity: 1967: Gal'perin, 1968; 1969: *GP,* 1971:1, p. 1; 1970: *GP,* 1971:7; 1972: *GP,* 1973:5, p. 3.

Volumes injected and withdrawn: GP, 1970:4, 1971:12, 1971:3, and 1975:1; *Gazovoe delo,* 1972:4.

TABLE 21. COMPARATIVE PIPE SIZE DISTRIBUTION FOR U.S. AND SOVIET GAS PIPELINE NETWORKS

Length and diameter of lines	USSR[a]		
	1964	1966	1967
Total length of gas lines (thous. km)	33.0	41.8	47.4
Diameter class in inches (percent on Jan. 1)			
10 and under	9.4	9.6	10.3
10.1–15	14.0	13.6	12.1
15.1–20	26.3	24.1	23.9
20.1–30	28.3	25.0	22.9
32	10.2	9.5	9.2
40	11.2	17.8	21.2
48	0.0	0.0	0.0

Note: Soviet pipe sizes have been converted from millimeters to inches for comparative purposes.

Sources: USSR: 1964–69: *GP,* 1970:4; 1970 and 1971: Rubinov, 1972, p. 32; 1972: ANSSSR, 1974, p. 260.

U.S.: Federal Power Commission, *Statistics of Interstate Natural Gas Pipeline Companies,* 1972.

States less than 1 percent of the total was 40 inches or larger. Table 21 shows no 56-inch (1,420 mm) gas pipeline, as there was none in operation as of the last year shown in the table. In 1974 the first line using 56-inch pipe (part of the Medvezh'e–Center system) was put into operation, and it was forecast that by the end of 1975 there would be about 3.7 thousand kilometers of such line in operation, which would be about 4 percent of the total system (*Stroitel'stvo truboprovodov,* 1974:5, p. 9 and *Soviet News,* November 12, 1974).

This increase in the length of the network, together with the large diameter, has made it possible to handle big increases in the amount of gas. Total throughput has grown from 113 billion cubic meters in 1965 to 231 billion cubic meters in 1973. In connection with the adverse shift in location, the average length of haul has increased appreciably, so that the total volume of transport work has increased even more. Between 1965 and 1970, the average length of haul rose from 650 kilometers to 917 kilometers. Although I have been unable to find this indicator for later years, the growing importance of the Turkmen, Vuktyl, and West Siberian fields has surely pushed the average up still further. It is expected that by 1980 the average length of haul will reach 2,230 kilometers (*GP*, 1975:1, p. 5).

But the capacity of the pipeline system has apparently not kept up with production capacity. There is shut-in capacity in some fields, and industry spokesmen attribute the failure to meet output goals in part to limitations in the ability of the pipelines to deliver the gas existing wells

USSR[a]					U.S.
1968	1969	1970	1971	1972	1972
52.6	56.1	63.2	67.5	72.5	255.7
9.0	10.2	11.0	10.7	11.0	26.7
12.2	11.3	31.8	31.9	31.5	8.0
23.2	22.4				16.2
22.2	21.2	27.4	26.5	25.9	41.3
8.3	8.5				7.2[b]
24.7	24.1	23.9	23.5	23.4	0.6[c]
0.0	0.9	4.2	5.7	6.1	0.0

[a] USSR distribution adds to less than 100 percent because some special purpose lines, such as condensate lines, are omitted from the distribution by size.
[b] 30″–36″.
[c] 40″–42″.

could produce. There are many reasons for this: inadequate treatment of gas to be shipped, resulting in the condensation of liquids that obstructs flow in the lines; unreliable equipment in compressor stations; pipe and control equipment inadequate to Siberian conditions, leading to breakdowns; and others. The problem with compressor stations is one of long standing. The Russians have been very slow about getting compressor stations finished on schedule, and lines reach their planned capacity only after long delays. One of the central problems is the compressors themselves. Soviet policy has emphasized gas burning turbocompressors of very large capacity—individual units of 4, 6, and 10 kilowatts. These have not been produced at the required rates, and those that have been installed are said to lack sufficient reliability, so that there are frequent breakdowns. In the early sixties, the gap was filled in part by using gas-motor piston compressors and centrifugal compressors that use electric motors as prime movers. In the late sixties, the share of gas burning turbocompressors increased significantly and they now account for over half of all installed compressor capacity. But the Minister of the Gas Industry, S. A. Orudzhev, has recently expressed the view that the only solution to the problem of reliability will be to shift to electric motors as prime movers (*GP*, 1975:1, p. 6). An alternative prime mover which has attractive features is the aviation turbine. Such units require less elaborate installation and would solve some of the transport and construction problems that have slowed the completion of compressor stations, especially in Siberia and the North. The Russians have been experimenting with a unit consisting of a turbo-prop engine (retired from aviation service) connected to one of the standard centrifugal compressors of 4,000-kilowatt capacity. Three of these have been in operation since 1972 in a compressor station in the Ukraine (*GP*, 1974:3). It is also reported that the Russians are interested in Rolls-Royce aviation turbines for this purpose (Moscow Narodny Bank, *Press Bulletin*, Oct. 16, 1974, p. 3). They are also testing one American-made unit (*GP*, 1975:3, p. 41), and have enlisted the aid of the Czechs in getting more reliable equipment (*GP*, 1975:2, p. 36). In short, the problem of reliable, large capacity, easily installed compressors seems likely to remain one on which they will need technology from abroad for a long time to come.

The proximate problem in pipelines is to master the construction of lines of still larger diameter than the 1,020-mm lines that now constitute the major part of the long distance network or the 1,420-mm lines recently commissioned. The volume of the flows from the giant West Siberian fields is to be such that several lines even of these larger diameters would be required to handle it. Big savings have been postulated from going to much larger diameter lines; in particular, there has been discussion of 2.5-meter lines, though it seems that experimental work on lines this large is no longer being seriously pursued.

Storage

Seasonal fluctuations in gas consumption create a need for significant storage capacity in market areas. Storage makes it possible to match the capacity of lines to average rather than peak flows, resulting in large capital savings. The longer the pipeline, the more advantageous storage becomes relative to increasing throughput capacity. The Russians do not have the depleted oil reservoirs near market areas which have provided cheap and convenient storage in the United States, and have substituted artificial underground reservoirs in appropriate formations. (For further details see *ESOG*.) But these have been rather expensive, and progress in developing and using them has been slow. Goals for capacity have been underfulfilled, and have had to be rescheduled several times. At one point it was planned to have a storage capacity of 17–18 billion cubic meters in operation by the end of 1970 (*SSSR v novoi piatiletke*, 1966, p. 31), but the goal for January 1, 1976, was later set at only 7 billion cubic meters. The latter now appears to be within reach, with capacity on January 1, 1971, of over 6 billion cubic meters, and with an actual injection in 1974 of 10.4 billion cubic meters (*GP*, 1975:1).

Consumption

The allocation of gas by sector of use is shown in table 22. The table apparently includes manufactured as well as natural gas, but excludes liquefied gases. Soviet sources have been very uninformative for years since 1970, perhaps because they are sensitive about the net import of gas, which grew from 0.3 billion cubic meters in 1970 to 3.5 in 1971, 5.9 in 1972, and 4.6 billion cubic meters in 1973.

An alternative classification of interest is a functional one (given in Umanskii and Umanskii, 1974) as follows for 1970.

Use	*Percent*
Industrial processing (kilns, furnaces, etc.)	33.2
Boiler fuel (except electric stations)	20.3
Electric stations	28.9
Household use (except in boilers)	5.6
Nonfuel	7.2
Export and own needs	4.8

Soviet consumption of gas continues to be characterized by a distinctive pattern in which industrial use for boiler and furnace fuels plays a much

Table 22. Consumption of Gas by Sector

(billion cubic meters)

Sector	1965	1966	1967	1968	1969	1970	1973
Household and municipal	15.3	16.1	18.6	21.1	23.6	26.2	30.0
Industry[a]	73.4	82.5	90.2	99.9	105.9	118.5	n.a.
Chemicals	6.1	7.7	9.1	10.7	11.2	13.0	n.a.
Metallurgy	18.4	22.4	25.0	27.4	29.4	32.6	n.a.
Machinery and metal-working	12.8	14.8	14.9	16.2	17.5	18.9	71.8
Oil and gas	12.7	13.4	14.0	16.2	16.6	19.3	n.a.
Construction materials and construction	13.9	14.9	16.9	18.1	18.8	20.7	68.3
Light industry	2.0	1.7	1.9	2.2	2.2	2.5	n.a.
Food industry	4.2	4.6	5.1	5.7	6.0	6.8	n.a.
Other industry	3.3	3.0	3.3	3.4	4.2	4.7	n.a.
Electric power stations	35.7	40.8	43.6	44.1	47.0	48.6	n.a.
Transport	0.4	0.4	0.6	0.6	0.6	0.9	n.a.
Agriculture	0.2	0.3	0.2	0.4	0.5	0.7	n.a.
Net export	0.4	0.8	1.3	1.7	2.7	−0.3	−4.6
Unexplained	2.8	3.0	3.4	2.8	1.2	4.3	n.a.
Total	128.2	143.9	157.9	170.6	181.5	198.9	242.0

Notes: n.a. = not available. The figures include manufactured as well as natural gas, but exclude liquefied gases.

Sources: Based essentially on *GP,* 1971:3, except for the following: 1973, households, machinery industry, and construction materials: *GP,* 1974:10, p. 2; 1973 total: production corrected for net exports.

[a] Including nonutility electric stations, which used 7.4 billion cubic meters in 1965 (*GP,* 1973:3) and 12.3 billion cubic meters in 1970 (*GP,* 1971:3).

larger role, and household use a much smaller role, than in the United States. Particularly important is the use of gas for electric power generation, amounting at the present time to about one-third of all gas. One market long neglected—agriculture and rural households—has now been given a high priority, and significant amounts are now going for this purpose.

As explained in *ESOG*, the Soviet consumption pattern is based on several features of the Soviet setting. Because of low incomes, Soviet households are smaller users of fuel and energy in general than U.S. households; much of the household space heating demand is met by by-product heat from heat–electricity combines in the big cities; planners have emphasized those uses where very large amounts can be absorbed with relatively small investments in distribution facilities.

Perhaps the most interesting question regarding the utilization of gas concerns the possibilities for substituting natural gas for other fuels, especially oil. Natural gas resources are much less tight than oil, and there are advantages to exporting oil rather than natural gas. Oil transport is cheaper, and involves fewer fixed commitments, for example. Given the very large amount of residual fuel oil now used as boiler and furnace fuel (estimated in an earlier section as 80–85 million tons) there would seem to be a very large potential for substituting gas for residual fuel oil. This would involve numerous other changes at the same time, especially in the mix of refinery outputs. Gas at present also plays so significant a role in electric power generation because of heavy use in the summer months when it is not needed for other uses and can be substituted for more expensive fuels in electric power generation. Further substitution would have to involve much more year round use, which depends on considerable progress in overcoming the storage and transportation bottlenecks. But when there is so large a market, now being served by residual fuel oil, it does seem that the Russians would find it advantageous to make this substitution to free oil for export.

Natural Gas Liquids and Liquefied Gases

An important subsector in the oil and gas industry is the processing of various gases to recover the hydrocarbons with molecular weights above methane, such as propane, butane, etc. The gas in some reservoirs contains large amounts of these components, which, at the high pressures in the reservoir, are in gaseous form, but condense when the gas is produced. The dissolved gas produced along with oil (which the Russians call *poputnyi gaz* and which we will describe as oil-well gas) is commonly rich in ethane, propane, and butane. Stabilization of crude oil in the field may also give rise to a light liquid fraction which contains considerable ethane, propane, and butane along with natural gasoline. Finally, the gases produced in refining processes, especially cracking, contain large amounts of ethane, propane, and butane. Because the raw materials occur in so many different circumstances and because there is a lot of latitude in deciding whether to recover them, and at what point and in what mixtures, this is a sector that is difficult to define. It is essentially gas processing, producing such liquid products as condensate and liquefied gases together with more or less pure methane suitable for pipeline transmission. These outputs can be utilized in various ways—as intermediates for further processing into chemical and fuel products, or directly as fuel. Gas processing involves all three parts of the industry—oil production, refining, and gas production—which, in the USSR, puts it in several administrative jurisdictions. Hence, it is treated vaguely in Soviet statistics and is a difficult sector to study. Statistics on the most consistently reported magnitudes are shown in table 23.

Until recently the Russians have tended to neglect gas processing and the recovery of natural gas liquids and liquefied gases. A large share of the oil-well gas has been flared rather than used in any way. As shown in table 18, the share of oil-well gas utilized dropped appreciably from about 70 percent in 1965 to 59 percent in 1974, which meant that the absolute amount flared rose from about 7 to 19 billion cubic meters in the same period. Of the oil-well gas utilized, only part is processed to recover the liquefiable components—less than half in the early seventies.

Soviet condensate resources are very large, but have so far not been extensively utilized. In 1965, 65 percent of natural gas production was from fields with appreciable amounts of condensate (Mel'nikov [ed.], 1968, p. 461), and the new fields that have subsequently become important—especially Orenburg and Vuktyl—are particularly rich in condensate. It is estimated (*GP*, 1971:7) that total condensate resources amount to 4.5 billion tons and that there are an additional 3.5 billion tons of other liquids (presumably the liquefiable fractions). Only a fraction of the condensate available from gas produced is extracted: the potential in 1970 is said to have been 10 million tons compared with the 3.8 million tons actually recovered (Mingareev and Luzin, 1972, p. 110). When withdrawal of gas from condensate reservoirs lowers formation pressure enough, condensation takes place within the reservoir and the condensate is lost. This loss can be prevented by recycling processed gas back into the reservoir, to be finally produced only after most of the condensate has been recovered. The Russians, however, have so far not used this approach, and have presumably already irretrievably lost large amounts of condensate.

TABLE 23. INDICATORS OF GAS PROCESSING, NATURAL GAS LIQUIDS, AND LIQUEFIED GASES

Year	Production of condensate (million tons)	Gas processed (billion cubic meters)			Production and consumption of liquefied gases (thousand tons)						Other liquid products in gas processing plants (thous. tons)
					Production[a]			Consumption			
		Gas-well gas	Oil-well gas	Total	In petroleum refineries	In gas processing plants	Total	Households	Petrochemical industry	Other	
1958	0.4	n.a.	n.a.	n.a.	195[b]	111[b]	308	n.a.	n.a.	n.a.	n.a.
1959	0.4	[c]	2.1	2.1	299	204	503	80	430	n.a.	n.a.
1960	0.7	n.a.	n.a.	n.a.	347	313	660	126	n.a.	n.a.	n.a.
1961	0.7	n.a.	n.a.	n.a.	508	442	950	n.a.	n.a.	n.a.	n.a.
1962	0.8	n.a.	n.a.	n.a.	689	567	1,256	282	889	85	n.a.
1963	0.8	n.a.	n.a.	n.a.	1,032	629	1,661	410	1,170	81	n.a.
1964	1.0	n.a.	n.a.	n.a.	1,379	920	2,299	n.a.	n.a.	n.a.	n.a.
1965	1.2	2.7	5.7	8.4	1,624	1,169	2,793	723	1,748	322	690
1966	1.4	2.9	6.6	9.5	1,819	1,308	3,127	920	1,963	300	861
1967	n.a.	2.9	7.5	10.4	2,002	1,522	3,524	1,140	2,200	256	918
1968	1.7	2.5	8.4	10.9	2,366	1,781	4,147	1,533	2,462	[d]	1,014
1969	2.3	3.4	9.7	13.1	2,685	1,733	4,418	1,849	2,462	[d]	1,792
1970	4.2	3.2	11.2	14.4	2,834	1,817	4,651	2,061	2,590	[d]	2,206
1971	5.3	3.4	12.1	15.5	3,215	2,000	5,215	2,160	2,670	385	n.a.
1972	6.6	n.a.	n.a.	n.a.	3,379	2,160	5,539	n.a.	n.a.	n.a.	n.a.
1973	7.6	n.a.	n.a.	n.a.	n.a.	n.a.	6,550	3,020	3,020	510	n.a.
1975 Plan	9.0	23.9	18.5	42.4	n.a.	n.a.	n.a.	n.a.	n.a.	n.a.	n.a.

(Continued)

Finally, oil refineries are run in a way that minimizes the production of liquefiable gases: the refinery gas that is produced is to a considerable extent burned directly for refinery fuel rather than being processed to produce liquefied gases.

Soviet fuel industry officials recognize the need to make better use of these potentials, and plans have always proposed much larger outputs than those finally achieved. The original 1970 output target for liquefied gases was 9 million tons compared with the 4.6 million tons actually produced, for example. The problem seems to be that as a fragmented activity intimately tied to the operations of several ministries but primary to none, gas processing is neglected. The gas processing plants, originally under the jurisdiction of the natural gas industry, were recently transferred to the oil industry, but the history of construction delays and failure to operate at designed capacity has continued without much improvement. This is mentioned as one of the serious failures in a review of oil industry operations in 1974 (*EG*, 1975:3, p. 2).

Despite these problems, the output of condensate and of liquefied gases has now reached significant amounts, and with the prominence of the Vuktyl and Orenburg fields, condensate output will be much larger very soon. How best to use these products does not yet seem to be fully resolved. Some condensate is processed into finished products

---

Notes to table 23.

n.a. = not available.

Sources: Condensate production: 1958: (output of Karadag field in Azerbaidzhan—the only significant condensate field at the time), Abramov, 1971, p. 87; other years: table 9.

Gas processed: 1959: Kortunov, 1967, p. 193; 1965–68: *Gazovoe delo,* 1970:3, p. 49; 1969–70: Mingareev and Luzin, 1972, pp. 108, 114 and *Gazovoe delo,* 1970:8, p. 12; 1971: *GP,* 1971:1; 1975 Plan: Baibakov, 1972 and *GP,* 1971:3.

Production of liquefied gases: Figures for 1958–65 are derived from *NIG,* 1967:1, p. 111, Kortunov, 1967, p. 194; and *NIG,* 1967:1, p. 111; 1966–69: *Gazovoe delo,* 1971:1, p. 40; 1970: *Stroitel'stvo truboprovodov,* 1971:3, p. 15, and Mingareev and Luzin, 1972, p. 114; 1971: *GP,* 1971:1, p. 2 and 1971:11 and Rachevskii, *et al.,* 1974, p. 4; 1972: Lisichkin, 1974-b, p. 48 and *GP,* 1972:3, p. 2; 1967, p. 195; 1971 nad 1973: Rachevskii, *et al.,* 1974.

Consumption of liquefied gases: Data is drawn from numerous sources, full of inconsistencies, especially for early years, when large amounts were reported as going to "other." It appears from other sources that most of that undistributed consumption was actually in petrochemicals. Major sources are *GP,* 1967:10, insert; *GP,* 1964:1, facing p. 29; *Stroitel'stvo truboprovodov,* 1971:3, Kortunov, 1967, p. 195; 1971 and 1973: Rachevskii, *et al.,* 1974.

Other liquid products: 1965, 1969–70: Mingareev and Luzin, 1972, pp. 108, 114; 1966–68: *Gazovoe delo,* 1970:3, p. 51.

[a] The figures in this table for production of liquefied gases differ appreciably from those in *ESOG*. This table is based on later, and more exact statements.

[b] Does not add to production total, because taken from different sources.

[c] Negligible.

[d] Included in households and petrochemical.

in the gas processing plants; much of it is run to stills with crude oil in traditional refineries. One suggestion is that condensate be "methanized" to help cover winter deficits, or to supply areas that cannot easily be served by natural gas pipelines.

The consumption of liquefied gases so far has been overwhelmingly in the chemical industry, and to a certain extent in household heating—especially in less densely settled regions and in rural areas that do not justify pipelines. Petrochemical production still takes over half of liquefied gas output but its share is declining in favor of more household use. This utilization pattern is rather different from that in the United States where there is a greater variety of uses, and chemical uses are less important in the total. It appears from a table in Kortunov, 1967 (he was formerly Minister of the Gas Industry), that auto transport was to be a growing use, but because output has grown so much less than planned, this use has not developed strongly.

A breakdown (in *GP*, 1972:12, p. 7 and 1970:10) different from that in table 23 shows 480,000 tons in 1969 and 665,000 tons in 1971 going to rural areas. This suggests that little progress has yet been made in the "gasification" of agriculture with liquefied gases. There seems to be an appreciable and growing export of liquefied gases, mostly to France. Liquefied gas is not systematically reported in the foreign trade statistics, but one source shows that 224,000 tons were exported in 1973 (*GP*, 1974:5, p. 4).

Natural Gas Exports

The issue of greatest current interest in regard to Soviet natural gas is the USSR's export potential. There are several possible directions in which increased gas exports could go—more gas to Eastern Europe to replace some of the oil the Russians are now sending there; increases in the amounts of gas to Western Europe already contracted for; and gas exports to Japan and to the United States in the form of liquefied methane. The last is probably a more complicated alternative than exports of gas to Eastern and Western Europe, since it involves much more novel technology for liquefaction and transport of the methane gas. Exports to the United States and Japan also depend on development of the more difficult Western Siberian fields. The North Star project, which the Russians have been negotiating with an American consortium, for example, would utilize the Urengoi field in Western Siberia, already discovered and explored, but located farther north than the Siberian fields now being developed (primarily Medvezh'e) and posing difficult problems of development and transport. The project for sending gas eastward to Japan and the U.S. west coast would be based on resources in Eastern Siberia that are more speculative as to size, but which also involve a serious transport problem. The feasibility or desirability of these projects involve many issues beyond the scope of this study, such as U.S. policy on credits and technology transfer, dependence versus independence, and others. Accordingly we will not try to discuss their desirability here. Rather it seems more useful to extract from the preceding discussion of the Soviet gas industry two important points. First, there seems no doubt that the USSR does have the gas to support such exports. There may be many problems in turning these resources into output, and the magnitude of the particular reserves on which the exports to Japan are to be based need to be verified. It is also no doubt wise to consider the tightness of the energy situation in the USSR in the future. But the question of whether the USSR can deliver those amounts of gas cannot be answered by projections of past patterns of output and domestic consumption.

The second conclusion from the preceding survey of the industry, relevant to cooperative ventures in developing Soviet gas, is that the Russians do need technical help very badly. That is an important bargaining counter that should be used by the United States to extract its full share of the mutual advantage to be had from any such deal. Soviet spokesmen often say that they are contemplating these projects for our good, rather than for themselves, and for that reason it is appropriate that the United States supply credit. The above discussion, however, suggests that argument is hardly valid. The westward movement of Siberian gas is of vital importance to the USSR, and the experience gained from U.S. cooperation in building the pipeline, or any spare capacity it might have, will be extremely valuable to the Russians.

8

ECONOMIC REFORM IN THE OIL AND GAS INDUSTRY

The period since 1965 has brought some significant changes in economic analysis and in the use of economic instruments in administration of the Soviet economy. The early expectation, engendered by the reform measures introduced in 1965–67 by Khrushchev's successors, that the Soviet economy would be reorganized along more decentralized lines with dependence on profit as the criterion for decision making, has not been borne out. Still, a number of the changes that were made have had great significance for the oil and gas industry. The most interesting ones are in the areas of pricing, economic modeling, and administrative reorganization.

Pricing and Costing

Several general principles introduced in the price reform of 1967 had special significance to the oil and gas industry. The acceptance by Soviet policy makers of an interest cost to be included in prices has been important for the oil and gas industry because of its high capital intensity. Similarly, the oil and gas industry is the major area of application of the rent charge which was accepted as part of the price reform. Also at this time the idea of a finding charge for natural resources was accepted, and such charges were introduced for oil and gas. The introduction of all these changes led to a fairly drastic overhaul of the whole pricing system for oil and gas.

To cover the new rent, interest, and finding charges the general level of crude prices had to be raised significantly. The price reform set eight regional prices for crude oil ranging from 8 rubles to 22 rubles, with each producer in a region receiving the zonal price. An exception is made for several areas (Azerbaidzhan, Western Kazakhstan, Sakhalin and Western Siberia) in which the regional administrative organ varies the price paid to the various producing units. These zonal prices are high enough to cover payment of interest on the depreciated value of fixed assets at 6 percent. They also cover the estimated finding costs, calculated as an average for the region over a recent period per ton of reserves explored, and charged at this rate per ton produced. Because of cost variations within regions, some producers earn high profits at these prices, and rental payments are levied individually on each producing unit to extract these windfalls.

The rent payments do not coincide with standard theoretical ideas about rent and were not intended or designed as measures for guiding enterprise-level decision making. Since the rent charge is fixed on a per-ton basis, it cannot help in guiding the enterprise to an optimal decision as to the internal margin, that is, how much to intensify production by measures to increase the ultimate recovery coefficient or to accelerate the extraction of oil from the field it is exploiting. Economic theory would suggest that if the price of oil is set at its value, and profit is made the success indicator for enterprise performance, the enterprise would then carry intensification to the point where marginal cost equals the price, which would be the optimum output from a national economic point of view as well. But if rent is being charged per ton of output, marginal revenue is made equal to the price less the rent charge, and this would inhibit a profit-maximizing enterprise from intensifying output to the appropriate level. In fact, however, the reform did not really intend to put the output decision in the hands of the enterprise or to make profit the sole success indicator. The payoff formula introduced in the oil industry as part of the reform was a function of output increases as well as of profit. Thus the combination of a new price, rent charges and profit indicators was actually intended only to extract differential income, and not to guide local decisions regarding output level. But since the rent charges seem to have been varied rather frequently and arbitrarily among different producers within a combine, they have not been very effective in eliminating income attributable to natural factors, and thus have not made income a reliable function of enterprise performance (*PKh*, 1974:11, p. 87).

The fact that the new prices for crude oil were much above the level before the reform is alleged to have had positive effects. It made it easier to interest producers in intensifying production and conservation and to motivate refiners toward more economical use of crude oil and more rational choices. For example, when oil was cheap, refineries found natural gasoline from gas processing plants more expensive than comparable fractions they could distill from crude oil, and so were reluctant to utilize natural gasoline. It is also said that the new prices have helped overcome the notion that exporting crude was more advantageous than exporting oil products, a notion based on the exaggeratedly low domestic cost of crude.

There is still a large gap between the price levels of crude oil and the prices of oil products, and a large turnover tax is still collected on the latter. It is said that the price reform shifted only about one-fourth of this markup back to the crude production stage in the form of rent.

The reform raised petroleum product prices considerably. The criterion generally suggested for setting the price of residual fuel oil was to make it more or less comparable to the price of coal (which was raised considerably itself), allowing for differential advantages in use. Apparently, however, residual fuel oil was in fact given a palpable price advantage compared with coal, except in some major coal producing regions such as Siberia and Kazakhstan (Torbin, 1974, p. 175). This was no doubt desirable in the seventies as consistent with the policy goal of shifting many boiler and furnace uses from coal to oil, but this principle probably needed review in the new post-1973 situation. I have been unable to find complete information on the new product prices set in 1967, but information on a fairly representative sample of products (shown in table 24) suggests that the price structure for products was changed considerably. Comparison with the prices of the early sixties (see appendix table A-22) shows that the price of gasoline was lowered in relation to other automotive and tractor fuels and the prices of gasoline, kerosene, and diesel fuel were lowered in relation to those of residual fuel oil. The reform also reduced to three the number of zones for zonal prices for products.

Natural gas prices were overhauled along the same lines as oil prices. Field prices were raised to permit the introduction of a finding charge, a capital charge, and rent, while the prices to final consumers were also raised to put gas on a level comparable with other principal fuels. The national average extraction cost for natural gas, as the Russians figure it, is less than 1 ruble per thousand cubic meters, but with significant variation among fields and regions. (Data on the cost of extraction in individual fields is available in numerous sources, such as Urinson *et al.*, 1973, p. 67.) The new field prices for gas introduced in 1967 exhibited extensive differentiation among zones and there was corresponding wide variation in rent charges. Beginning January 1, 1971, however, field prices paid to producers were simplified to two prices—5 rubles per thousand cubic meters for Uzbek, Turkmen, and West Kazakhstan gas, and 6 rubles per thousand cubic meters everywhere else. This is high enough to cover the cost of the highest-cost producer in each region, including a flat finding charge of 1 ruble per thousand cubic meters uniform for all areas, and a capital charge. The differences in net income among the different producers are then smoothed out by a rent

TABLE 24. PRICES FOR SELECTED PETROLEUM PRODUCTS, INTRODUCED IN 1967 (rubles per ton)

Product	Zones		
	I	II	III
Aviation gasoline B-70	145.00	153.00	165.00
Automotive gasoline			
A-76	97.50	104.00	110.50
A-72	91.00	97.50	104.00
A-70	84.50	91.00	97.50
A-66	78.00	84.50	91.00
A-93	123.50	130.00	136.50
Kerosene, tractor	35.00	39.00	44.00
Auto–tractor diesel fuel			
Summer	60.00	64.00	69.00
Summer, sulfur content not more than 0.5%	66.00	70.00	75.00
Winter	66.00	70.00	75.00
Automotive lube oil with special additives	190–230	200–240	210–250
Regular	135.00	145.00	155.00
Residual fuel oil, grades 100–400			
Sulfur less than 0.5%	25.00	28.00	32.00
Sulfur up to 2%	23.50	26.50	30.50
Sulfur up to 3.5%	22.50	25.50	29.50
Fuel oil, marine	26.00	29.00	34.00
Road asphalt	25.00	29.00	35.00

Sources: Spravochnik ekonomista kolkhoza i sovkhoza (Handbook for State and Collective Farm Economists), Moscow, 1970, pp. 408–412; Avrukh, 1970, p. 133; and Torbin, 1974, p. 93. An explanation of the coverage of the different zones is given in the first source. The general scheme is that Zone III covers the Far East, Zone II most of Eastern Siberia plus some of Western Siberia, while all the rest of the RSFSR is in Zone I. According to Umanskii and Umanskii, 1974, p. 507, all republics other than the RSFSR are in Zone I.

charge ranging from 0.5 to 4 rubles, also levied per thousand cubic meters (Umanskii and Umanskii, 1974, p. 508). Gas is sold to final consumers at zonally differentiated prices ranging from 11 to 24 rubles per thousand cubic meters. The difference between these prices and the field price covers transport cost (around 2 rubles in the early seventies for the average haul then existing) and a discount to the city distribution organizations (2.5 to 3 rubles), but the prices must still generate a considerable profit and turnover tax as well.

As in the case of oil, the new pricing system for gas is said to have provided a strong incentive for economizing on the use of gas, and for making better decisions within the fuel sector regarding such choices as primary energy sources for compressor stations, use of gas for refinery fuel, and so on.

Fuel Allocation Models and Shadow Prices

The new prices fixed for oil and gas are not expected to carry the full burden of guiding economic decision making. It is a distinctive feature of the Soviet system that project makers designing new facilities (in this connection those that will be using large amounts of fuel, such as electric power stations) often base their decisions on cost indicators other than actual transaction prices. One of the novelties of the last several years in Soviet fuel planning is the gradual development of what seem to be rather elaborate and carefully constructed optimizing models to generate shadow prices (the *zamykaiushchie zatraty*) for the various fuels in various regions. An interesting example is a set of shadow prices for several years up to 1980 generated by an optimizing model which determined the breakdown of regional and interfuel output, the interfuel composition and interuser allocation within each consuming region, and the interregional linkage of production and consumption (Melent'ev, 1973). These show a close correspondence to the pattern of zonal differences embodied in the gas prices set by the 1967 reform and thus provide corroboration of the rationality of the latter. There is one interesting difference, however. The shadow prices were computed separately for the heating season and the slack summer demand period, and the resulting differentials are much larger than the 10 percent discount for summer use of gas that was established in 1967 and is embodied in current prices.

The oil and gas industry is full of situations suitable for the use of operations research methods of all kinds, and such methods seem to have been introduced widely. They are especially important in pipeline design decisions and in the operation of refineries. There are now a large number of computer installations in the industry, and the Ministry of the Oil Industry, at least, is busily engaged in the creation of a ministry-wide system of data collection and retrieval based on a computer purchased abroad.

Administrative Reform

The oil and gas industry has experienced frequent reorganization since 1965, much of it involving reshuffling of production organizations among administrative jurisdictions. Examples are the shift of responsibility for construction of pipelines from the Ministry of the Gas Industry to a separate ministry, the shift of responsibility for gas processing from the gas industry to the oil industry, and the authorization to the Ministry of the Gas Industry to develop its own drilling program rather than depending on the Ministry of Geology and the

Ministry of the Oil Industry. Many of these changes represent simply a continuation of the kind of organizational flux and the bureaucratic power game described in *ESOG*. They are interesting as an illustration of the general problem of organization for economic administration and as a case study of how economic reform has worked in practice. But they are generally peripheral to the concerns of this study, and I will remark on only two points.

1. The general trend evident since about 1970 has been a considerable centralizing movement, with the elimination of intermediate levels of administration and a trend toward a functional rather than area basis of organization. In the oil industry the main organization below the ministry remains the regional combine (*ob'edinenie*), such as Bashneft' or Glavtiumenneftegaz, of which there are twenty-three. Within the combines there has been considerable centralization, so that specialized services such as drilling, geophysics, and repair, and such functions as finance and supply, are controlled by combine headquarters rather than by the field-based producing units. In the gas industry a similar consolidation began in 1974, which eliminated by merger about 200 of the 580 primary enterprises in the ministry. The chain of command between the ministry and these new larger organs was also shortened by removal of intermediate links. This centralizing trend is tied up with the economy-wide movement to create computer-based management systems, which the Russians call ASU or ASUP.

2. An interesting novelty in both the oil and gas ministries is a reorganization of the research and development network designed to make it more effective in the generation of technical progress. All the regional research and project-making organs have been consolidated into a system of regional institutes (in the oil industry they now go by names like TatNIPIneft'—the Tatar Scientific Research and Project-making Institute for Oil) subordinated directly to the local regional combine. A large part of the research and development network remains centered in the ministries, in the so-called *golovnye* institutes, which are expected to supervise the work of the local research and development organizations. But it will be interesting to see if this does in fact remove some of the traditional barriers between research and production, thereby giving researchers greater access to producers and making them more responsive to production needs in improving technology.

9

INTERNATIONAL ECONOMIC RELATIONS

Soviet fuel policy has been conditioned in part by the goal of exporting energy to earn foreign exchange, especially hard currency. As indicated in chapter 1, net exports amount to about 10 percent of all primary energy production. In the case of oil, nearly one-third of total output serves export objectives rather than domestic needs (see table 25). Until recently, gas has played an unimportant role in trade in energy products, and, until 1973, the actual balance constituted a small net import. But in view of the relatively less restrictive resource situation for gas, the Russians see gas as the growing element in energy exports, and have contracted for large exports both to socialist countries and to Western Europe over the next decade or so. Deliveries to Western Europe are now beginning to grow fairly rapidly, and in 1973 (when exports reached 6.8 billion cubic meters) the direction of trade turned to a small export surplus. The Russians now have contracts with

TABLE 25. DISPOSITION OF SOVIET OIL OUTPUT BETWEEN EXPORT AND DOMESTIC USE

(million tons)

Year	Crude oil output	Exports in field equivalents[a]	Imports in field equivalents[a]	Net exports	Domestic use	Share exported (percent)
1965	242.9	71.7	2.0	69.7	173.2	29
1966	265.1	81.8	1.8	80.0	185.1	30
1967	288.1	87.5	1.6	85.9	202.2	30
1968	309.2	95.8	2.1	93.7	215.5	30
1969	328.4	100.6	3.8	96.8	231.6	29
1970	353.0	106.2	4.8	101.4	251.6	29
1971	377.1	116.3	7.0	109.3	267.8	29
1972	400.4	118.3	9.6	108.7	291.7	27
1973	429.0	130.6	15.6	115.0	314.0	27
1974	458.8	128.8	6.2	122.6	336.2	27

Source: Crude oil output from table 9.

[a] Products were first converted to a crude equivalent basis on the assumption that refinery and fuel expenditure and losses were 15 percent of the fuel refined; the sum of crude oil and products in crude equivalents was then converted to field equivalents on the basis of an assumed loss rate of 5 percent.

Western European countries to deliver about 23–25 billion cubic meters of natural gas per year by 1980. Additions are possible, if the tripartite West German–Soviet–Iranian deal works out,[1] and if the proposed sales to Sweden are agreed on.[2] It was also estimated that Soviet exports to Eastern Europe can be over 20 billion cubic meters by 1980,[3] but since those estimates were made, the rise in price may make it difficult for the Eastern European countries to buy this much. Possible exports of liquefied gas to the United States and Japan are one of the biggest items on the development agenda of the gas industry.

The development of oil exports in total and by geographic area is seen in table 26, and table 27 shows trends in unit values. Unfortunately, beginning in 1968, the Russians ceased to report information separately for crude oil and oil products by trade partner, so that some interesting issues remain obscure. Significant features of this trade are the following.

1. The year-to-year rate of growth of exports has been somewhat erratic, with no clear trend. There was a check to growth in the years from 1969 through 1972, but the Russians responded to high world prices with a renewed export expansion effort in 1973. In 1974 the total export of crude oil and products declined by 2 million tons, but there was a sharp decrease in imports from the Middle East, so that net exports rose by about 7 million tons.

2. The requirements of the socialist countries continue to place a heavy demand on Soviet export capacity. Their share of the total exports increased from about 44 percent in the mid-sixties to 62 percent in 1974. Especially important here are the Eastern European countries, which account for most of Soviet exports to socialist countries. Cuba is a significant and growing element in the total, however. The Eastern European countries are very poor in energy resources, although Rumania has oil and gas and Poland has large coal resources. Eastern European oil and gas output are shown in table 28. The one country with significant oil production—Rumania—has been unable to increase output significantly. Natural gas has grown rather more: by 75 percent between 1965 and 1974.

The really significant change in the oil situation in the socialist countries is the extraordinary expansion of Chinese output. The importance

[1] This tripartite deal would not be a drain on Soviet output; Soviet exports would be less than the imports from Iran (*Economist*, July 27, 1974).

[2] The pipeline delivering Finnish gas has been designed for a capacity of 3 billion m^3 per year, with the idea that it can handle gas deliveries to Sweden as well.

[3] Moscow Narodny Bank, *Press Bulletin*, June 19, 1974. The gas pipeline to Eastern Europe now being built with the resources of the East European recipients is to have a throughput capacity of 15.5 billion m^3 (*GP*, 1975:2, p. 35).

TABLE 26. SOVIET OIL EXPORTS BY PRODUCT AND GEOGRAPHIC AREA, 1966–74[a]
(million tons)

Product and destination	1966	1967	1968	1969
World total	73.6	79.0	86.2	90.8
Products[b]	23.3	24.9	27.0	26.9
Gasoline	2.7	3.3	3.4	3.3
Kerosene	1.3	1.2	1.3	1.5
Diesel fuel	8.3	8.7	10.1	9.9
Residual fuel oil	10.6	11.4	11.7	11.6
Crude	50.3	54.1	59.2	63.9
Socialist countries[c]	32.3	35.4	41.5	47.6
Crude only	25.5	27.3	n.a.	n.a.
Asia, total	0.8	1.0	1.3	1.4
Products only	760	973	n.a.	n.a.
Eastern Europe	25.3	27.6	32.4	37.9
Crude only	21.0	22.6	27.2	32.3
Cuba	5.1	5.3	5.3	5.8
Western Europe	29.9	35.6	38.2	37.1
Crude only	17.5	22.4	n.a.	n.a.
Rest of world	11.4	8.0	6.5	6.1

n.a. = not available.

Source: Soviet foreign trade handbooks, except exports of crude oil to Eastern Europe after 1967, based on Eastern European trade handbooks.

[a] The figures in this table represent actual exports and so differ from those in table 25, where actual exports were adjusted to a field-equivalent basis.

TABLE 27. PRICES OF OIL AND OIL PRODUCTS IN SOVIET TRADE
(prices in rubles per ton)

Item	1966	1967	1968
Exports			
Crude oil and products, Eastern Europe	16.4	16.0	15.7
Crude oil and products, Western Europe	10.1	10.8	11.5
Crude oil, all areas	11.8	11.9	12.1
Products, all areas	15.7	15.9	16.9
Ratio of product price to crude price (all areas)	1.33	1.34	1.40
Imports			
Products from Rumania	33.0	34.6	34.4
Crude oil, all areas	n.a.	n.a.	n.a.
Share of oil and products in value of all exports (percent)	12.0	11.9	12.3

n.a. = not available.

1970	1971	1972	1973	1974
95.8	105.1	107.0	118.3	116.2
29.0	30.3	30.8	33.0	35.6
3.5	4.1	4.5	5.5	5.8
2.1	2.4	2.4	2.3	2.6
11.4	11.4	12.4	14.2	15.8
11.4	11.9	11.1	10.4	10.8
66.8	74.8	76.2	85.3	80.6
50.4	55.4	60.2	67.7	71.7
n.a.	n.a.	n.a.	n.a.	n.a.
1.5	1.3	0.9	1.1	1.6
n.a.	n.a.	n.a.	n.a.	n.a.
40.3	44.8	48.9	55.3	58.7
34.4	38.5	44.4	49.8	
6.0	6.4	7.0	7.4	7.6
38.4	41.8	41.0	43.7	38.1
n.a.	n.a.	n.a.	n.a.	n.a.
7.0	7.9	5.8	6.9	6.4

[b] Total exceeds sum of components because some products are omitted.

[c] Exports to Albania and Yugoslavia included in total only, not in regional subtotals.

1969	1970	1971	1972	1973	1974
15.4	15.3	15.4	15.7	16.0	18.1
10.6	11.9	15.7	15.2	25.6	57.2
12.3	12.3	14.1	14.4	15.8	28.8
16.6	17.2	19.5	18.4	32.0	56.8
1.35	1.40	1.38	1.28	2.0	1.97
35.4	36.4	39.2	40.7	32.9	33.0
n.a.	8.9	11.1	14.3	16.7	66.1
11.7	11.5	13.3	13.1	15.2	20.9

Source: Soviet foreign trade handbooks.

TABLE 28. OUTPUT OF CRUDE OIL AND NATURAL GAS IN SOCIALIST COUNTRIES

Country	1966	1967	1968
Crude oil (thousand tons)			
Rumania	12,826	13,206	13,285
Hungary	1,706	1,686	1,807
Poland	400	450	475
Bulgaria	404	500	475
Czechoslovakia	190	200	205
China			
Yugoslavia	2,222	2,374	2,494
Natural gas (million m³)			
Rumania	18,046	19,857	21,036
Hungary	1,552	2,045	2,687
Poland	1,290	1,463	2,402
Czechoslovakia	1,070	1,017	1,108
Yugoslavia	n.a.	n.a.	n.a.

n.a. = not available.

Sources: Statisticheskii ezhegodnik stran-chlenov Soveta Ekonomicheskoi Vzaimopomoshchi (Statistical Yearbook of the Member Countries of the Council

of this development is not to make China independent of the USSR, since the USSR ceased supplying oil to China well before Chinese output began to grow. Rather, in a period when oil is an important element in international relations, it gives the Chinese a significant weapon to use against the Russians, whether in offering oil to the Japanese to dissuade them from helping the Russians develop Siberia, or in trying to induce a China slant in U.S. foreign policy. There is an interesting question whether China could use this weapon to increase her leverage in Eastern Europe, though that seems unlikely. Some believe that Chinese oil output could reach 100 million tons by 1976 and 400 million tons by 1980. (*The Economist*, January 18, 1975, p. 80). Such an output would permit very large exports in 1980, perhaps 100 million tons or more, and make China a more important oil exporter than the USSR.

3. There are some interesting price changes. At one time Eastern European countries were paying a higher price for crude oil and oil products than Western European countries, but that relationship is now reversed. The Russians have been able to take advantage of rising world market prices in their exports to capitalist countries but have been bound by the stable-price formula used in trade within the CEMA (Council for Mutual Economic Assistance) group. Prices for intra-CEMA trade were supposed to be fixed for the period through 1975, but the Russians have succeeded in breaking that formula, and negotiated a much higher price for 1975. It is reported (*The New York Times,* January 28, 1975) that the price has been fixed with the Hungarians at

1969	1970	1971	1972	1974
13,246	13,377	13,793	14,100	14,287
1,754	1,937	1,955	1,977	1,989
439	424	395	300	392
325	334	305	200	190
210	203	194	200	171
14,600	20,100	25,600	29,600	n.a.
2,699	2,854	2,967	3,200	n.a.
23,093	23,990	25,606	26,600	28,005
3,235	3,469	3,713	4,100	4,813
3,672	4,975	5,164	5,600	5,811
1,185	1,204	1,222	1,100	1,042
n.a.	1,100	1,200	1,200	n.a.

of Mutual Economic Assistance) and statistical handbooks of the respective countries; Yugoslavia and China from UN statistical yearbooks.

36 rubles per ton, and, to judge from past practice, the price will be uniform to all Eastern European countries. Since the Hungarians were paying approximately 16 rubles per ton for crude oil in 1973, this represents more than a doubling of the price. The outcome of this renegotiation effort is also a new formula for pricing during the five-year period after 1975. Recent CEMA practice has been for prices in each five-year period to be set as the average world market price in the preceding five-year period, but the Russians have understandably objected that this formula for oil is too adverse to their interests. The new formula employs a moving average under which prices on oil exports to Eastern Europe in each year will be based on the world market price averaged over the preceding five years. This will permit the high oil prices of the mid-seventies to influence CEMA prices sooner. The Russians have no doubt also pressed hard for an upward revision of the price for natural gas, but at the time of writing no announcement has been made on that score. They have an alternative lever in this case, through the requirements they are putting on the Eastern European countries to participate in financing the pipeline which will deliver the expanded flows of gas to Eastern Europe.

4. The ratio of oil products exports to crude exports has increased, gradually and consistently, presumably in an effort to increase the foreign exchange earning potential of the oil designated for export. Also interesting is the rise in Soviet imports of oil from the Middle East. It has increased significantly, though most of this oil probably never enters the USSR, but is shipped directly from the Middle East to Eastern

Europe or other export markets. Soviet trade statistics include such transactions in both imports and exports, even if the goods never cross the Soviet border. It is occasionally suggested that the Russians may be able to expand this kind of operation greatly, and thus share in the gains the oil producing countries are making from the energy crisis. In this connection it seems significant that the average price on Soviet purchases from the Middle East rose greatly in 1973 and again in 1974. This oil is obtained from the Middle Eastern countries (primarily Iraq) under various kinds of barter arrangements in return for past Soviet loans and assistance. Even so, the Middle Eastern countries have been able to push the price upward as world oil prices rise, suggesting serious limitations on the leverage the Russians are able to exert on the Middle Eastern countries in oil matters.

The USSR and Eastern Europe

The USSR is ambivalent about its relationship with Eastern Europe. It feels an obligation to supply oil to Eastern Europe for two reasons: a growing flow of energy is necessary to keep those economies growing; and, as the sole supplier of so vital a commodity, the USSR obtains a very powerful economic sanction enhancing its political power over these countries. On the other hand, oil is a very attractive candidate for export to hard currency areas, especially since the rise in oil prices. The commitment of oil to Eastern Europe cannot be requited by price increases alone, since the USSR needs hard currency to pay for the technologically advanced imports from western countries that play a very important role in current Soviet growth strategy. On the other hand, if the Russians decide to tell the Eastern Europeans to get their energy somewhere else, they contribute to weakening the bloc-preference approach they have heretofore encouraged, and also place these countries under a staggering burden of earning convertible foreign exchange.

The Soviet response to this dilemma is to encourage the Eastern Europeans to develop their own energy resources, and to seek oil and gas from Middle Eastern producers on a barter basis. The USSR has also strongly advocated that Eastern European countries base new electric generating facilities on nuclear fuel. By the seventies all of them were building significant amounts of such capacity, and some of these plants are now beginning to come into operation. For a long time the Russians were no doubt unwilling to share nuclear power plant technology with Eastern Europe, and, as the developers of the technology, must have sought to extract the largest possible economic advantage from sharing it. But they seem now to be encouraging nuclear power in Eastern Europe very strongly, and have recently created a new agency within CEMA, *Interatomproekt*, to provide design and technical services for building such plants. Finally, they have also demanded Eastern European participation in the expansion of Soviet oil and gas output. I have been unable to find details, but, according to a Soviet author, Czechoslovakia has for some time had an agreement with the USSR for joint development of oil, in which Czechoslovakia provided pipe and oil field equipment on credit, with the credit to be repaid by Soviet shipments of oil. East Germany has apparently had a similar agreement and Poland also recently signed such an agreement (Oleinik [ed.], 1972, p. 144; and *Soviet News*, December 17, 1974).

In the new situation since 1973, the Russians have pushed this line even harder. N. K. Baibakov, Chairman of the Gosplan, in announcing the 1975 Plan, said that during 1975, "For the first time on the territory of the USSR a number of objects will be built completely by the efforts and resources of construction organizations of the CEMA partners." He mentioned specifically the 2,800 kilometer gas line from Orenburg to the western border, which will supply gas to Eastern and Western Europe, and the construction of an oil pipeline from Polotsk in Belorussia to a new refinery being built at Mazheikiai in Lithuania (*EG*, 1974:52). The Eastern Europeans will supply not only the construction effort but also a hard currency credit to cover part of the cost of the pipe. These are very large projects, and the Russians' ability to shift them to the Eastern Europeans will free some very scarce Soviet capacities for other uses.

The Middle East

The relationship of Soviet oil policy to the Middle East also calls for comment. Soviet oil production cost is far above Middle Eastern cost, and it would be advantageous if the USSR could meet its needs with cheap oil from the Middle East rather than from domestic production. The resources the USSR is spending on expanding its own oil and gas industry, if invested in the Middle East instead, would give a much bigger payoff. Alternatively, it has been suggested that it would be a great gain to the Russians if they could shed the burden of supplying oil to Eastern Europe through shifting their Eastern European partners to Middle Eastern suppliers instead. The Russians could then export to the world market the 60 to 70 million tons of oil now being sent each year to Eastern Europe. But there is a fatal contradiction in these recipes for getting something for nothing—it is not in the interest of the Middle

Eastern producers to supply oil to the socialist world cheaply, especially not if the effect will be to permit the USSR to become a competing energy supplier in the world market.[4] The same general analysis holds for gas. Iran was originally willing to sell gas to the Russians at very low prices (about 18 cents per thousand cubic feet) because what was involved was associated gas that could not be used in Iran and was being flared. But with the advent of liquefied natural gas technology, this gas acquires a much greater value, and the Iranians have already insisted on and gained an increase in the price (to 57 cents per thousand cubic feet). The Russians have also agreed to an increase in the price paid for gas imported from Afghanistan. Cheap energy in the Middle East in a world of high energy prices constitutes as tempting a prize to the USSR as to anyone, but there is no way the Russians can cut themselves in on that wealth for nothing, except by political or military leverage.

The USSR and Eastern Europe ought to be able to take advantage of Middle Eastern oil income at second hand, however, in several ways. It is difficult for them to compete with western countries as a goods supplier, but arms offer a more tempting possibility. Secondly, it ought to be possible for Comecon to fashion some kind of asset in which the Middle Eastern countries could invest their hard currency earnings. There are interesting problems in designing this asset to make it tempting, but the possibility is there, as indicated by a recent agreement in which Kuwait lent Hungary $40 million (Moscow Narodny Bank, *Press Bulletin*, January 15, 1975).

In assessing the role of the USSR in the world energy markets, it is not only trade in these commodities that matters, but also Soviet participation in the effort to expand the production of energy outputs. What will be the Soviet stance in relation to exploration for and development of oil and gas, and in relation to supply of the inputs required for expansion—especially equipment, technology, and investment resources?

The USSR has played some role as provider of technology in the past. It has performed exploration, built refineries, and provided oil field

[4] Under certain assumptions some cartel members might be motivated to act as discriminating monopolists. If a producer within the cartel had unused production capacity and a strong desire for additional foreign exchange earnings, but could not increase its market share within the cartel, it might be willing to sell to the USSR or Eastern Europe at a concessionary price, assuming this would not undercut the cartel in its main markets. Iraq, to take a possible example, might be willing to expand sales to Eastern Europe at a special price in order to increase output, and on the calculation that Eastern Europe could supply the kind of investment goods and technical assistance needed for development. This transformation of Iraq underground oil assets into above-ground production assets would no doubt be advantageous at an oil price well below the present cartel price for oil.

equipment for underdeveloped countries. The USSR continues to negotiate such agreements, as in a recent one with Argentina, in which the Russians promised to supply technical help for oil exploration. But what they have to offer along this line is quite insignificant, and the concern now is with the flow of technology and assistance *to* the USSR. We have emphasized that they need help in all areas of the oil and gas industry—refining, production, exploration, transport, equipment for drilling deep wells, bits, offshore drilling equipment, and so forth. In addition to importing modern capital goods (such as submersible electric pumps and turbocompressor units for gas pipelines), the Russians are also seeking and getting closer forms of cooperation from U.S. firms, such as technical help with exploration and in offshore drilling, and in the North Star gas project. They would like to pay for an expansion of this flow through energy exports later. What the U.S. response should be is controversial, involving very broad issues such as the desirability of technology transfer, the rationale for subsidized export credits, and the issue of dependence versus independence in energy. Much of this has nothing to do with the Russian oil and gas industry per se, and is not, therefore, an appropriate topic for extensive discussion here. But it does seem clear enough from this survey that energy exports from the USSR *are* feasible, and that Soviet need for the technical assistance which the United States and other advanced capitalist countries could provide is very great. The outcome will not turn only on the economic evaluation each side makes of such trade, of course. On the Soviet side energy exports are related to a broader controversy over development strategy in general. Expansion of energy exports is part of a strategy that seeks growth through technology imports, and therefore must provide something to pay for them. But this strategy is a departure from the historic Soviet pattern, and no doubt some in the USSR see détente and the economic interdependence with the capitalist world that accompanies it as a mistake. In a world where raw materials are becoming scarcer, they would argue, it is better to conserve Soviet natural resources for the appreciation they will undergo and the long-term competitive advantage they can provide. This, incidentally, is an old theme in Russian debates on economic development, long antedating Soviet doctrine. In the nineteenth century, there was a long controversy over whether Russia should follow the western development path, and should seek foreign capital and technology. Those who wanted to do so saw natural resources, and often specifically Siberian resources, as the principal asset that could be used to pull Russia out of economic backwardness. The current policy of détente and the current economic strategy exhibits many fascinating parallels with the strategy of economic development pushed by Sergei Witte as Russian finance minister at the end of the nineteenth century. Witte's policy ultimately fell victim to a more reactionary, xenophobic, and militaristic line, and this could happen once again.[5] Indications of renewed controversy on this point are seen in an emphatic statement made in 1974 by V. D. Shashin, Minister of the Oil Industry, that the Soviet Union does not need western help in developing its oil and gas resources (later retracted),[6] in assertions that natural resources should not be squandered, and in a theme that has recently occurred in high level policy statements, with many subtle variations of the kind that keep Kremlinologists busy, about the need to take a "proprietary" attitude toward Soviet resources. Brezhnev first put this forward in a speech in fall 1974, and the formula has been repeatedly cited, and indirectly objected to or modified in many subsequent announcements by others. An editorial in *GNIG*, 1975:3 is very eloquent on the rationale for expansion of gas exports, but clearly on the defensive. Some of these signals have been subsequently reversed, but the point is that the line so prominently enunciated for several years to the effect that the Soviet Union could supply energy resources in exchange for credits and equipment is closely intertwined with a political line, and could be reversed.

[5] The nineteenth century experience is described in Theo von Laue, *Sergei Witte and the Industrialization of Russia,* New York, Columbia University Press, 1963. Rereading that book is very rewarding as a way of putting the present stage of Soviet development experience in perspective.

[6] *New York Times,* May 23, May 28, 1974.

10

CONCLUSION

The major part of this study has been concerned with the oil and gas industry proper, but the relationship of decisions about oil and gas to other aspects of energy policy has been evident at many points. Whether one's interest in Soviet oil and gas arises because of an interest in the prospects for Soviet participation in the world energy market, or out of a desire to compare the Soviet situation regarding oil and gas with our own, it is necessary to relate oil and gas developments to the framework of Soviet energy policy in general. Thus it seems appropriate in concluding this study to return to some of the broad issues of overall energy policy noted in the introduction.

For nearly two decades Soviet economic planners have relied on oil and gas almost exclusively as the means of satisfying the growing energy demands associated with economic growth. This choice has surely been a sensible one. First it has meant big savings both in investment requirements and in current production outlays. It is difficult to conceptualize and measure the capital costs of utilizing alternative primary energy sources, but, as a simpleminded illustration, the data on investment by industrial branch reported in the Soviet statistical handbooks show that investments of 15.7 billion rubles in the oil and gas industry during the Eighth Five Year Plan permitted output growth of about 240 million tons of standard fuel, whereas investments of 7.4 billion rubles in the coal industry achieved an increase in output of 20 million tons of standard fuel. These figures are far from exhaustive, omitting much ancillary investment in transport, exploration, and treatment required to utilize these primary outputs, and such outlays are probably greater for oil and gas than for coal.[1] Even so, the implication is that to have met the energy needs of growth by expansion of the coal industry would have involved a several-fold greater investment burden than did the oil and gas choice.

Second, the emphasis on oil contributed powerfully to solving the problem of trade expansion. In the decade beginning in the mid-fifties,

[1] See especially Luzin, 1974, for extensive data on investment in ancillary activities.

oil provided about 40 percent of the increment in export earnings. After 1965, oil export earnings grew less rapidly than all exports, but from 1973 on, rising prices have once again given oil exports a strikingly important role in the growth of export earnings. In 1973, with volume virtually unchanged, the increment in oil earnings constituted one-fourth of the total growth in foreign exchange earnings. In 1974 total oil and gas exports rose only slightly, but provided nearly half of the increment in foreign exchange earnings (*EG*, 1975:15, p. 20).

Finally, Soviet oil and gas expansion have helped the USSR's socialist partners in Eastern Europe to cope with their energy poverty. It seems doubtful that there is any way the Russians could have provided an equivalent amount of energy to Eastern Europe in other forms, or that these countries could have expanded domestic output enough to have provided the flow of energy inputs that has been vital in achieving the growth they have had.

In reflecting on how the Russians were able to implement this oil and gas option it strikes one that the technological requirements were not especially demanding. Throughout the industry Soviet producers were able to apply relatively familiar technology, which, even if it were not already mastered in the USSR, was familiar and widely diffused in the world economy. A significant job of adaptation was needed, and turbodrilling exemplifies the independent development of a major innovation. But in general the expansion of the oil and gas industry did not involve innovations near the frontier of technology. Much of Soviet technological policy in the oil and gas sector represented "extensive development," that is, replication of existing technology on a wider scale. For example, the expansion in the number of wells constructed was handled simply by using, on a greatly extended scale, the same kind of low-quality tubular goods that had always been used. In other cases the Russians were able to evade domestic roadblocks to upgrading technology by large-scale importation of more or less standard goods, as in the case of large-diameter pipe for gas lines, or submersible pumps. What are sometimes described as outstanding technical achievements—such as the construction of very long pipelines of very large average diameter or the extensive introduction of waterflood production techniques—reflect the scale of resources and output or institutional structure more than technological novelty or ingenuity. Indeed, some of these achievements should be construed as brute-force flanking movements around technical problems, much as the Soviet space program developed unusually powerful rockets to overcome the weight problem, rather than attempting to miniaturize. Thus, the large diameter of pipelines is in part a compensation for the fact that the strength of Soviet-made line pipe has limited working pressures to 55 atmospheres, which is below the design choices usual elsewhere. These statements are made not to belittle the Soviet achievement, but as background for a hypothesis that the USSR may now be facing a rather different situation that will require more novel technology. Examples are the need to make significant jumps in the size of compressors, or to employ mobile offshore rigs, which may move them much closer to the frontier of world experience in a number of technological areas.

In speculating on the future of the Soviet energy sector, I believe it is misleading to assert, as is sometimes done, that the Russians, too, have an energy crisis. On the other hand, there have been some changes in the technological and resource environment that pose obstacles to a continuation of the heavy reliance on oil and gas, and lead to a special Soviet version of the energy problem. The Soviet problem differs from that of the United States and Western Europe, since the USSR has the geological resources that should permit it to remain self-sufficient, an option which may simply not exist for Western Europe, or which would be extremely expensive for the United States. The dependence–independence issue for the USSR is how much to export rather than how much to import, but whatever point is chosen within the range of this variable, the USSR has more alternatives for meeting the associated level of output. Furthermore, the Soviet situation is distinctive in that a very large share of Soviet oil and gas is consumed in uses—especially as boiler and furnace fuel—where it could reasonably easily be replaced by coal. Liquid automotive fuel accounts for less than 14 percent of total Soviet energy consumption, but about 24 percent of U.S. energy consumption.[2]

On the other hand, the Soviet problem is reminiscent of the one the United States faces with respect to western coal—all Soviet alternatives involve a transport problem. The strategic choices to be made by Soviet energy policy makers in planning energy output are among alternative ways to move Siberian energy to the European areas or to export points. The transport problem might be sidestepped by heavy emphasis on nuclear power, or moderated by various breakthroughs that would cheapen or permit the production and transport of Siberian energy sources.

Mastering high-voltage, long-distance transmission of electric power

[2] These are approximate shares, calculated for the USSR as the ratio of consumption of gasoline, kerosene, and diesel fuel (table 15) to total energy consumption (table 1) in 1970 and for the United States as total petroleum product use in transportation as a share of total energy consumption in 1972 (Federal Energy Administration, 1974, p. 46).

would make it possible to use any of the Siberian primary sources—coal, hydropower, or gas—to meet incremental needs or provide some relief in oil demand. Soviet planning in energy research and development may indeed be heading in that direction. In a report at the end of 1974, the chairman of the State Committee on Science and Technology, V. A. Kirillin, suggested that primary emphasis should henceforth be placed on coal rather than on oil and gas for electric power generation (*Kommunist,* 1975:1, pp. 43–51). But, given the coal situation in European Russia, this means resort to Siberian coal, primarily from the Ekibastuz and Kansk-Achinsk basins. Since these are respectively 2,400 and 4,000 km from the central economic region (in which Moscow is located), such a policy seems to require solution of the power-by-wire problem. Kirillin devoted a part of his report to this problem, and the general perspective of the Soviet establishment on this matter is perhaps revealed in a long article in *Pravda* by Academician Popkov about the time of the Kirillin speech, which is very sanguine on the prospects for efficient long-distance power transmission (see *Current Digest of the Soviet Press,* Vol. XXVI, Nos. 41, 48, 1974).

On its face, the use of Siberian gas would seem to be an easier answer. Gas would seem to have so many advantages over coal-by-wire from Kansk-Achinsk that it is puzzling why so much attention is being given to coal and long-distance transmission. The usual suggestion is that the problem is transport cost and reliability. In fact, however, a gas pipeline bringing Tiumen' gas to the central region[3] has been completed, and gas from Siberia is moving there in significant quantities. Perhaps experience with this line is showing it to be so costly or unreliable that Soviet gas planners are wary of building more such lines, especially in areas still farther north. Perhaps there is a desire to preserve gas for export to the United States or Western Europe, or to Eastern Europe to save the oil now being exported there.

Other obvious possibilities, such as improved methods of transporting coal itself, production of liquid fuel from coal for transport westward, or conversion of Siberian gas to methanol for easier transport, seem to have been slighted in Soviet discussions of future energy policy.

In summary, the future of Soviet energy policy, including choices as to the volume of export and the emphasis given to oil and gas in output and export, seems closely connected to the choices Soviet planners make regarding technology and energy research and development. They are

[3] This line goes from Nadym and Punga via a southerly route through Nizhniaia Tura and Perm, and is now being doubled. Another line is also being constructed along another more northerly route from the Tiumen' fields via Nadym to Ukhta, where it will feed gas into the line that takes Vuktyl gas to the central region.

now in the process of making important decisions about research and development priorities in this sector, and, as one aspect of the problem, deciding how much they want to rely on technical borrowing from abroad. Soviet energy policy is now undergoing a transition, and the most productive way to get some perspective on the future of Soviet oil and gas would be more intensive study of the kind of bets the Soviet planners are making about the prospects for innovation that would enhance the attractiveness of some alternative fuel policies over others.

APPENDIX

This appendix is composed of tables, photographed directly from *ESOG*, which have been updated in this book. To facilitate easy reference to comparable tables, the following list was prepared.

TABLE A-1. SOVIET FUEL AND ENERGY PRODUCTION, TRADE, AND APPARENT CONSUMPTION

(million tons of conventional fuel)

Year	Production										Net trade[a] net exports (−) net imports (+)	Apparent consumption incl. cent. wood
	Mineral fuels						Firewood					
	Coal	Oil	Gas	Peat	Shale	Total	Centralized	Comprehensive	Hydropower	Total energy incl. cent. wood		
1928[b]	29.8	16.6	0.4	2.1	n.a.	48.9[c]	5.7	52.9[c]	0.2	54.8	−3.9	50.9
1932	54.0	30.5	1.4	5.4	0.1	91.4	14.4	(60)	0.5[d]	106.3	−10.3	96.0
1933[b]	64.0	30.6	n.a.	5.5	n.a.	103.7[c]	16.6	63.9[c]	0.8	n.a.	−8.7	n.a.
1937[a]	108.4	40.8	3.0	9.8	0.2	162.2	25.2	(80)	2.8	190.2	−3.8	136.4
1940	140.5	44.5	4.4	13.6	0.6	203.6	34.1	84.0[c]	3.3	241.0	+2.3	243.3
1945	115.0	27.8	4.2	9.2	0.4	156.6	28.4	n.a.	3.0	188.0	n.a.	n.a.
1950	205.7	54.2	7.3	14.8	1.3	283.3	27.9	62.6[c]	7.5	318.7	+10.0	328.7
1951	221.9	60.4	7.9	16.3	1.6	308.1	29.8	n.a.	7.3	n.a.	n.a.	n.a.
1952	236.9	67.7	8.1	15.3	1.8	329.8	28.4	n.a.	7.8	n.a.	n.a.	n.a.
1953	252.3	75.5	8.7	15.8	2.1	354.4	29.8	n.a.	9.8	n.a.	n.a.	n.a.
1954	275.3	84.8	9.5	18.5	2.4	390.5	32.7	n.a.	9.2	n.a.	n.a.	n.a.
1955	310.8	101.2	11.4	20.8	3.3	447.5	32.4	62.5[d]	12.3	492.4	−2.1	490.1
1956	325.1	119.8	15.2	18.4	3.5	482.0	32.0	n.a.	14.8	528.8	−7.5	521.2
1957	351.7	140.6	23.2	22.5	3.7	541.7	32.9	n.a.	19.6	594.2	−18.8	575.4
1958	362.1	161.9	33.9	21.1	4.5	583.5	32.9	56.7[c]	22.5	n.a.	−27.4	638.9
1959	370.0	185.3	42.5	23.0	4.6	625.4	34.0	n.a.	22.7	682.1	−38.2	643.9
1960	373.1	211.4	54.4	20.4	4.8	664.1	28.7	60.0[e]	23.8	716.6	−51.1	665.5

1961	370.1	237.5	70.8	19.6	5.2	703.2	26.0	59.0[e]	27.1	756.3	−65.7	690.6
1962	379.3	266.1	85.9	13.3	5.8	150.4	29.4	n.a.	32.2	812.0	−77.1	734.9
1963	388.4	294.7	105.1	21.7	6.5	816.4	30.7	n.a.	33.2	880.3	−89.1	791.2
1964	403.3	319.8	127.0	22.2	7.1	879.4	32.8	n.a.	33.1	945.3	−100.2	845.1
1965	420.2	347.3	151.3	22.0	7.5	948.3	32.0	60.0	35.5	1,015.8	−103.5	912.3
1965 (SYP goal)[d]	441.0	343.2	177.9	27.2	7.3	996.6	18.6	n.a.	37.4	1,052.6	n.a.	n.a.
1970 plan	487.0	500.5	271.4	34.3	9.8	1,303.0	32.0	n.a.	36.0	1,371.0	n.a.	n.a.
1980 plan[d]	850.0	1,000.0	850.0	[f]	50.0[f]	2,750.0	[f]	n.a.	200.0[g]	2,950.0	n.a.	n.a.

n.a. = not available or not applicable.

[a] For mineral fuels, by conversion of output in natural units on the basis of 1940 conversion ratios.

[b] For mineral fuels, by conversion of output in natural units on the basis of 1932 conversion ratios.

[c] Shimkin, 1962.

[d] Melent'ev *et al.*, 1962, pp. 38, 173.

[e] N. G. Poleshchuk, *Osnovnye voprosy toplivno-energeticheskoi bazy SSSR*, M, 1965, p. 34.

[f] Firewood and peat are included with shale.

[g] Includes atomic energy.

Fuel output: Except as noted, data are taken directly from recent statistical handbooks, or are based on output in natural units, converted to conventional tons by actual Soviet statements about average heat content, or by analogy with nearby years.

Hydropower: There are many divergent statements in Soviet sources for the energy equivalents of hydroelectric power, but I have tried to choose figures to conform to the concept of actual hydroelectric energy output converted at central station fuel rates for the corresponding year. I have not, however, been able to find a completely consistent series on central station fuel rates. The Russians have generally converted hydropower to fuel equivalents by using the fuel rate for central steam stations for the corresponding year. There is little to quarrel with here, except that (1) it would probably be better to use the average rate for all stations rather than for central stations alone (the difference between the two is far from negligible) and (2) in figuring the central station rate the Russians attribute part of fuel consumption to by-product heat output rather than to electricity output. The share contributed by hydropower is thus somewhat understated.

Net trade is figured from detailed data on trade in physical units, converted to conventional tons by the usual conversion ratios, except that for coal and coke, one natural ton is treated as equal to one conventional ton. Up to 1940, all solid fuels are included in this correction; since then, only coal and coke among the solid fuels. Gas is included only since 1955. There has been no appreciable export of electric power.

Firewood: Centralized wood, the official series, includes only firewood produced by the lumbering industry. The comprehensive series includes rough estimates for firewood produced by the population for its own use. The comprehensive firewood figures for 1932 and 1937 are interpolated estimates. The figure for 1965 is implied in *EG*, 1966:37, p. 5.

TABLE A-2. REGIONAL DISTRIBUTION OF PRIMARY FUEL PRODUCTION AND CONSUMPTION, 1960

Region	Output		Consumption		Surplus (+) or deficit (−) conv. fuel (*million tons*)
	Per cent of total	Conv. fuel (*million tons*)	Per cent of total	Conv. fuel (*million tons*)	
Northwest, and European North	3.5	24.0	6.8	43.7	−19.7
Center	4.7	32.4	16.0	102.7	−70.3
Volga	17.0	117.6	6.1	39.1	+78.5
North Caucasus	9.1	62.9	4.7	30.1	+32.8
Ural	12.1	83.7	16.3	104.6	−20.9
Western Siberia	11.1	76.7	6.4	41.1	+35.6
Eastern Siberia	3.8	26.5	4.8	30.8	−4.3
Far East	2.4	16.4	2.9	18.6	−2.2
Ukraine	24.3	168.2	22.3	143.1	+25.1
West	0.8	5.5	1.9	12.2	−6.7
Transcaucasus	5.0	34.4	2.8	18.0	+16.4
Central Asia	2.3	15.6	2.4	15.4	+0.2
Kazakhstan	3.6	25.1	4.6	29.5	−4.4
Belorussian S.S.R.	0.5	3.8	1.6	10.3	−6.5
Moldavian S.S.R.	0.0	0.0	0.4	2.6	−2.6
Total	99.9	692.8	100.0	641.8	51.0

Note: The percentage distribution of consumption by region is from *VS*, 1962:5, p. 51, and originated with the 1960 fuel balance tables prepared by the Russians. Therefore, it must not include wood prepared by population, and probably excludes hydro energy. There is a possibility that it does not include all light petroleum products. Applying the percentage figures to the apparent consumption of fuels, excluding decentralized firewood, as shown in Table 1, gives the absolute figures shown for consumption.

Sources: The distribution of output is based on calculations from basic output data for individual fuels and heat content figures.

TABLE A-3. EXPLORATORY WORK IN THE SOVIET OIL AND GAS INDUSTRY

Item	Unit	1940	1946	1950	1955	1958	1963	1965
All exploratory work[a]	million rubles	n.a.	54.64	241.36	345.05	480.2	1,061	1,369
All exploratory drilling	million rubles	n.a.	37.68	187.04	235.90	311.8	750	1,027
Geological and Geophysical	million rubles	7.1	16.96	54.32	109.15	168.4	311	369
Geophysical	million rubles	3.3	4.0	17.6	n.a.	70.5	200	129
Seismic	% of geophysical	40	n.a.	n.a.	65[b]	65	78[b]	n.a.
All exploratory drilling[c]	1,000 meters	531	651	2,127	2,242	3,369	5,316	5,568
Prospecting drilling	% of exploratory	n.a.	n.a.	n.a.	40	48.8	52.4	46.0
Core drilling	1,000 meters	221	214	1,127	2,187	3,041	3,600	n.a.
Geophysical crews[d]	number	86	107	286	429	673	n.a.	n.a.
Seismic	number	18	24	118	247	432	871	n.a.
Granimetric	number	21	30	52	107	143	n.a.	n.a.
Magnetic	number	16	16	21	4	6	n.a.	n.a.
Electrometric	number	31	37	88	68	92	n.a.	n.a.
Well logging	million meters	n.a.	n.a.	n.a.	27.7	n.a.	n.a.	45[e]

[a] "All exploratory work" aggregates the total volume of exploratory work in constant price terms, i.e., the estimate prices (*smetnaia stoimost'*) calculated in terms of constant norms and prices, at which work is shown in the project lists; and this is divided into several components as shown. Underlying these value indices is a whole family of physical indicators of volume. For exploratory drilling the measure is meters drilled. For geological and geophysical work the main indicators used in aggregative planning and in any general discussion are core drilling footage and the number of geophysical parties working during the year. The meaning of the core drilling series is imprecise—the numbers cited in different sources vary widely. Core drilling is usually distinguished from deep exploratory drilling by its purpose (to map structures rather than look for oil), by the equipment used, and by the lesser depth and diameter of the hole. However, some core drilling goes down to about 2,000 feet, and much core is taken in exploratory wells.

[b] Approximate.

[c] "All exploratory drilling" refers to the series described as *razvedochnoe burenie*. The wells included under this heading are officially subdivided into prospecting (*poiskovyi*) wells, outlining (*okonturivaiushchii*) wells, and study (*otsenochnyi*) wells drilled to obtain additional geological information. The distinctions are not actually observed in reporting practice, but in recent years the Russians have provided figures on prospecting drilling as a separate category within all exploratory drilling, though drilling footage can be given in this breakdown only approximately on the basis of estimates.

[d] Two cautions are in order concerning the number of geophysical crews. (1) In addition to the number of crews doing geophysical prospecting (called *polevye geofizicheskie* crews) the Russians often refer to a larger concept of geophysical work, which includes the crews that do well logging and other kinds of physical measurements, perforations, and so on, on wells at the development stage. The latter are called *promyslo-geofizicheskie* crews. (2) The Russians are very vague about the definition of the *polevye geofizicheskie* numbers, but they clearly refer not to the annual average number in operation, but to the total number of crews sent into the field at some time during the year. They are thus not comparable with the figures based on crew months compiled by the Society of Exploration Geophysicists for the United States. Quite an adjustment is necessary to eliminate this difference. In 1954 the average length of the field season was 5.4 months in seismic work, 6.0 months in gravity work, and 6.8 months in electric work (*GNIG*, 1958:2). Assuming that these periods are also applicable to 1955, the 429 crews shown in the table performed only 2,463 crew-months of work. The source further says that about 15 per cent of field time was idle time. This adjustment reduces the number of crews defined as an annual average from 429 to 173.

[e] Data for 1964.

Sources: Major sources include: Keller, 1958, pp. 11, 12, 13, 46; Buialov, 1960, pp. 65–67; Brenner, 1962, pp. 82, 89; Mirchink *et al.*, 1965, pp. 94, 95; Brenner, 1960, p. 58; Lisichkin, 1958, p. 36; and *Tekhnologiia i tekhnika bureniia neftianykh i gazovykh skvazhin*, p. 3. The remaining information comes essentially from the periodic annual reviews of exploration performance and plans in such journals as *GNIG*, *NIG*, *NKh*.

TABLE A-4. INDEX OF GROWTH OF SOVIET OIL RESERVES (A + B), JANUARY 1

January 1	Index
1940	80
1946	100.0
1947	112.0
1948	137.5
1949	149.5
1950	164.0
1951	197.7
1952	333.0
1953	371.6
1954	423.3
1955	477.6
1956	510.0
1957	556
1959	598
1961	765
1965	897
1966 (plan)	1,017
1966 (actual)	[a]

[a] Reserves on January 1, 1966, are reported as "more than 1.5 times January 1, 1959" in *GNIG*, 1966:6, p. 2, which suggests that there was no increase in 1965.

Sources: 1946 through 1957, Keller, 1958, pp. 5 and 6; a link between 1940 and 1955 is given in *Opyt razrabotki neftianykh mestorozhdenii*, Moscow, 1957, pp. 30–1. Statements in NKh, 1959:9, p. 2, and 1961:10, p. 5, supply links to 1959 and 1961, and a statement in *GNIG*, 1965:4, p. 7, to 1965. The goal for the Seven Year Plan was to raise reserves by 1.7 times, and this figure is presumably the ratio between January 1, 1959, and January 1, 1966.

TABLE A-5. TIME BUDGETS FOR ALL DRILLING, 1950, 1958, AND 1962

(*minutes per meter drilled*)

	1950	1958	1962
Productive time	79.5	37.0	42.4
Rotating bit	25.5	10.9	13.4
Raising and lowering tools	17.5	7.2	9.3
Running and setting casing	8.8	3.7	3.8
Auxiliary work	27.7	15.1	15.9
Non-productive time	57.5	20.0	22.6
Stoppages for repair	10.4	3.8	5.0
Geological complications	7.7	3.7	4.0
Breakdowns	14.7	4.9	5.2
Organizational stoppages	24.8	7.7	8.3
Total	137.0	57.0	65.0

Source: Time budgets in percentage terms for these years are available in *Prom SSSR*, 1964, p. 222. These have been converted into minutes-per-meter on the basis of the minutes of rig time per meter implied by the commercial speed indicator.

TABLE A-6. EXPENDITURES OF MAJOR MATERIALS IN DRILLING

Item	1940	1951	1955	1957	1963
Drill pipe (*kg/m*)	13.6	12.0	10.0	8.4	7.9
Casing (*kg/m*)	71.3	51.2	50.4	46.3	43.6
Cement (*kg/m*)	48.7	48.3	43.6	42.1	38.0
Total drilled (*1,000 meters*)	1,947	4,707	5,012	6,161	9,148
Total expenditures:					
Drill pipe (*1,000 tons*)	26.6	56.5	50.1	51.8	72.3
Casing (*1,000 tons*)	138.8	241.0	252.6	285.2	398.8
Cement (*1,000 tons*)	94.6	227.4	218.5	259.4	347.6

Sources: Dubrovina (ed.), *Neftianaia promyshlennost' SSSR*, p. 98; Lisichkin, *Ocherki razvitiia neftedobyvaiushchei promyshlennosti SSSR*, p. 401; *NKh*, 1961:1, p. 4; Umanskii, 1965, p. 216.

Table A-7. Selected Indicators of Soviet Drilling Operations

Item	1940	1946	1950	1951	1952	1953	1954	1955	1956	1957
Meters drilled (*thousands*)	1,947	1,229	4,283	4,707	4,905	5,362	4,940	5,012	5,090	6,161
Exploratory	531	651	2,127	2,375	2,279	2,482	2,268	2,242	2,315	2,868
Development	1,416	578	2,156	2,332	2,626	2,880	2,672	2,770	2,775	3,293
Turbodrilling (*per cent of total*)	2.4	6.5	23.7	30.6	41.4	53.6	65	83.1	87.6	n.a.
Drilling with electric drills (*thousand meters*)	neg.	neg.	11.1	10.2	20.6	41.6	44.7	48.7	49.2	71.2
Drilling with bits 9″ and smaller (*thousand meters*)	neg.	neg.	neg.	neg.	n.a.	n.a.	n.a.	200	n.a.	211
Exploratory drilling for gas (*thousand meters*)	n.a.	n.a.	n.a.	n.a.	n.a.	n.a.	n.a.	321	343	555
Commercial speed (*m/rig/mo*)										
Exploratory	233	177	209	205	197	216	230	306	342	401
Development	412	372	629	652	675	711	749	893	940	1,082
Mechanical speed (*m/hr*)		1.98	2.38							
Exploratory	1.26	1.52	1.68	1.81	2.17	2.70	2.68	3.19	4.37	5.19
Development	2.08	2.68	3.92	3.69	4.87	6.05	6.70	8.76	9.75	10.62
Hole drilled per bit (*meters*)										
Exploratory	19.6	n.a.	23.5	22.3	20.5	21.4	20.3	20.8	21.0	19.1
Development	33.4	n.a.	36.7	36.3	37.2	33.8	32.2	29.5	28.9	30.8
Cost of wells (*rubles/m*)										
Exploratory	n.a.	87.6	118.2	113.9	124.3	125.6	137.5	124.0	111.1	106.9
Development	n.a.	47.7	45.6	45.3	45.5	46.8	50.3	48.5	45.9	45.6
Number of wells completed	1,812	950	2,893	3,081	3,482	3,451	3,496	3,320	3,449	3,461
Exploratory	359	342	1,074	1,219	1,372	1,337	1,421	1,394	1,156	1,388
Development	1,453	608	1,819	1,862	2,110	2,114	2,075	1,926	2,293	2,073
Average depth of wells (*meters*)										
Exploratory	1,108	1,120	1,349	1,421	1,497	1,530	1,626	1,822	1,866	1,845
Development	990	1,021	1,146	1,160	1,254	1,307	1,350	1,454	1,449	1,478
Number of rigs operating	469	430	1,119	1,246	1,270	1,277	1,103	852	799	838
Exploratory	187	302	837	952	950	944	810	602	556	588
Development	282	128	282	294	320	333	295	255	243	250

Item	1958	1959	1960	1961	1962	1963	1964	1965	1965 (SYP goal)
Meters drilled (*thousands*)	6,887	7,148	7,715	8,363	8,873	9,148	10,003	10,754	16,000
Exploratory	3,369	3,762	4,023	4,533	4,808	4,861	5,316	5,614	10,400
Development	3,518	3,386	3,692	3,830	4,064	4,287	4,687	5,140	5,600
Turbodrilling (*per cent of total*)	85	86.5	83.0	83.0	83.4	82.2	80	n.a.	60–65
Drilling with electric drills (*thousand meters*)	54	80	(300)[a]	n.a.	n.a.	174	n.a.	n.a.	2,000
Drilling with bits 9″ and smaller (*thousand meters*)	211	783	1,711	2,834	3,451	4,053	4,812	5,324	10,500
Exploratory drilling for gas (*thousand meters*)	746	918	1,132	1,500	1,618	1,734	1,804	1,937	2,468

Commercial speed (*m/rig/mo*)									
Exploratory	416	419	401	397	390	368	374	378	780
Development	1,084	996	993	966	991	991	1,049	1,090	1,760
Mechanical speed (*m/hr*)									
Exploratory	5.24	5.23	4.88	n.a.	4.35	4.1	n.a.	n.a.	n.a.
Development	10.59	9.85	9.20	n.a.	8.45	8.7	n.a.	n.a.	n.a.
Hole drilled per bit (*meters*)									
Exploratory	18.8	18.7	19.9	n.a.	16.9	16.6	n.a.	n.a.	n.a.
Development	30.0	28.1	28.5	n.a.	25.6	25.9	n.a.	n.a.	n.a.
Cost of wells (*rubles/m*)									
Exploratory	108.3	105.7	111.9	n.a.	n.a.	141.6	142.7	n.a.	82.0
Development	46.5	47.3	49.4	n.a.	n.a.	53.3	55.9	n.a.	35.9
Number of wells completed	3,810	3,779	3,892	n.a.	4,298	n.a.	n.a.	n.a.	n.a.
Exploratory	1,517	1,706	1,660	1,856	1,932	n.a.	2,156	n.a.	n.a.
Development	2,293	2,073	2,232	n.a.	2,366	n.a.	n.a.	n.a.	n.a.
Average depth of wells (*meters*)									
Exploratory	1,857	1,862	1,928	1,995	2,063	2,060	2,094	2,195	2.360
Development	1,535	1,567	1,586	1,620	1,614	1,627	1,621	1,647	1,820
Number of rigs operating	933	1,017	1,130	1,282	1,369	1,460	1,556	1,631	1,358
Exploratory	666	738	824	952	1,027	1,100	1,184	1,238	1,096
Development	267	279	306	330	342	360	372	393	262

Sources: Meters drilled: Nar khoz SSSR, various years, except 1965 from *GNIG*, 1965:6, p. 2, and *EG*, 1966:9, p. 11.

Turbodrilling as per cent of total: 1946: *Vsesoiuznoe soveshchanie novatorov neftianikov*, M-L, 1951, p. 22; 1940, 1950, 1955–56: *NKh*, 1957:11, p. 32; 1951–54: Gurevich *et al.*, 1958, p. 104; 1958: L'vov and Keller, 1960, p. 66; 1959: *OGJ*, Sept. 14, 1959; 1960 and 1965 plan: *Tekhnika i tekhnologiia bureniia glubokikh skvazhin*, p. 18; 1961: Alexandrov, p. 7; 1962: *Prom SSSR*, 1964, p. 227; 1963: *NKh*, 1964:9–10, p. 101; 1964: *NKh*, 1965:1, p. 2.

Drilling with electric drills: L'vov and Keller, 1960, p. 71; Fomenko, 1961; and Umanskii, 1965, p. 42.

Drilling with bits 9″ and smaller: 1955: *NKh*, 1961:10, p. 6; 1957–58: Serenko, p. 29; 1959–62: *Prom SSSR*, 1964, p. 227; 1963: *Pravda*, December 21, 1965; 1964–65: *Nar khoz SSSR*, 1965, p. 177.

Exploratory drilling for gas: 1955–57: Smyshliaeva, 1961, p. 76; 1959–60: Brenner, 1962, p. 83; 1958: Bokserman, 1964, p. 19; 1959–64: *GNIG*, 1965:11, p. 3; 1965: *GP*, 1967:5.

Commercial speed: Nar khoz SSSR, various years, and *Prom SSSR*, except for 1946, which is from Keller, 1958, p. 32.

Mechanical speed: 1940: Brenner, 1960, p. 58; 1946–50: Vsesoiuznoe soveshchanie neftianikov, *Burovye dolota*, M-L, 1952, p. 98; 1951–53: Ministerstvo neftianoi promyshlennosti, *Burovye dolota*, M, 1955, p. 4; 1954–56: *Neftianaia promyshlennost' v 1959–65 gg.* M, 1958, p. 10; 1957–60: *Tekhnika i tekhnologiia bureniia glubokikh skvazhin*, 1962, p. 29; 1962–63: Titkov, 1965, p. 11.

Hole drilled per bit: NKh, 1954:1, p. 17 and 1955:2, p. 9; Gurevich *et al.*, p. 98; Lisichkin, 1958, p. 65; Ministerstvo neftianoi promyshlennosti SSSR, tekhnicheskoe upravlenie, *Tekhnologiia i tekhnika bureniia neftianykh i gazovykh skvazhin*, p. 5; *EG*, June 27, 1964; *Normirovanie raskhoda*, 1959, p. 46; *Tekhnika i tekhnologiia bureniia glubokikh skvazhin*, 1962, p. 29; and Titkov, 1965, p. 11.

Cost of wells: 1950–56: Keller, 1958, p. 41; 1957–58, 1965 plan: Zasiadko, 1959, p. 63; 1946: based on data in Keller, 1958, pp. 26, 41, 44; 1959–60: *NKh*, 1961:10, p. 18; 1963: *EG*, June 27, 1964; 1964: *PKh*, 1965:11, p. 34.

Number of wells completed: 1946: Dunaev, I, 1957, p. 65, and Brenner, 1960, p. 58; 1940, 1950, 1955, 1958–60: Kalamkarov, 1961, p. 30; 1951–54: Dunaev, I, 1957, p. 65, and Buialov, 1960, p. 67; 1956–57: Buialov, 1960, p. 67; 1962: *EG*, June 1, 1963 and Avrov, 1963, p. 35; 1961: Avrov, 1963, p. 35; 1964: *GNIG*, 1965:4, p. 1.

Average depth of wells: 1940: Lisichkin, 1958, p. 53; 1946: L'vov and Keller, 1960, and Lisichkin, 1958, p. 53; 1950: Gurevich *et al.*, 1958, p. 97; 1951–56: GOSINTI, *Neftianaia i gazovaia promyshlennost' v 1959–65 gg.* M, 1958, p. 7; 1951–62: *Prom SSSR*, 1964, p. 221; 1963: Umanskii, 1965, p. 22; 1964: *PKh*, 1965:11, p. 34; 1965: *NKh*, 1966:9, p. 5; 1965 plan: Kalamkarov, 1961, p. 31.

Number of rigs operating: these are the average annual number implied by the figures for footage and commercial speed.

[a] Planned figure.

TABLE A-8. GROWTH OF SOVIET OIL OUTPUT

Year	Million tons	Year	Million tons
1901	12.0	1951	42.3
1913	10.3	1952	47.3
1926	8.3	1953	52.8
1928	11.6	1954	59.3
1932	21.4	1955	70.8
1937	28.5	1956	83.8
1938	30.2	1957	98.3
1939	30.3	1958	113.2
1940	31.1	1959	129.6
1941	33.0	1960	147.9
1942	22.1	1961	166,0
1943	18.1	1962	186.2
1944	18.3	1963	206.1
1945	19.4	1964	223.6
1946	21.7	1965	242.9
1947	26.0	1966	265.0
1948	29.2	1967 plan	286.2
1949	33.4	1970 plan	345–355
1950	37.9	1980 plan	690–710

Sources: Nar khoz SSSR, various issues; 1941: *NKh*, 1965:5, p. 1; 1942–44: Shigalin, 1960, p. 113; 1966 and 1967 plan: *NKh*, 1967:1, pp. 1 and 3; 1970: *Pravda*, April 10, 1966. Up through the middle of 1962, the goal for 1970 was described as 390 million tons: *NKh*, 1962:6, p. 1; 1980: Table 1.

TABLE A-9. SOVIET CRUDE OIL OUTPUT, BY REGIONS, SELECTED YEARS, 1930–65

(*thousand tons*)

Region	1930	1937	1945	1950	1955	1960	1965
R.S.F.S.R.	7,445	5,746	5,675	18,231	49,263	118,861	199,929
Volga-Ural	6	1,003	2,833	10,990	41,220	105,600	173,629
Tatar A.S.S.R.	—	—	10	1,020	14,600	43,360	76,449
Bashkir A.S.S.R.	—	962	1,300	5,250	14,200	26,700	43,907
Kuibyshev oblast'	—	17	1,020	3,480	7,250	21,750	33,382
Stalingrad oblast'	—	—	n.a.	n.a.	2,300	5,200	6,182
Saratov oblast'	—	—	n.a.	440	1,900	3,600	1,341
Orenburg oblast'	—	—	n.a.	n.a.	510	1,600	2,623
Perm oblast'	6	24	190	300	570	2,300	9,745
Western Siberia	—	—	—	—	—	—	1,000
North Caucasus	7,344	4,346	n.a.	6,310	6,540	12,100	20,700
Krasnodar krai	n.a.	1,433	700	3,000	3,890	6,540	6,200
Chechen-Ingush A.S.S.R.	n.a.	2,744	890	2,500	2,120	3,610	8,976
Dagestan A.S.S.R.	n.a.	169	550	920	520	300	1,001
Stavropol'skii krai	—	—	n.a.	—	neg.	2,700	4,554
Sakhalin	95	355	n.a.	620	950	1,500	2,411
Komi A.S.S.R.	—	42	n.a.	330	550	820	2,223
Belorussian S.S.R.	—	—	—	—	—	—	39
Azerbaidzhan S.S.R.	10,596	21,414	11,541	14,822	15,305	17,833	21,500
Ukraine S.S.R.	—	neg.	250	293	531	2,159	7,580
Turkmen S.S.R.	12	452	629	2,021	3,126	5,278	9,636
Fergana Valley	48	390	517	1,409	1,128	2,084	2,152
Georgian S.S.R.	—	9	36	43	43	34	30
Kazakh S.S.R.	349	490	788	1,059	1,397	1,610	2,022
All U.S.S.R.	18,451	28,501	19,436	37,878	70,793	147,859	242,888

Notes to Table A-9

— indicates "none."

Note: Most zeros on the right are not significant.

Sources: 1930: Glavnoe upravlenie neftianoi promyshlennosti, *Neftianaia promyshlennost' S.S.S.R. v tsifrakh,* Moscow-Leningrad, 1935, p. 14.

1937: The distribution by republic is given in postwar statistical handbooks, but not allocation with R.S.F.S.R. Volga-Ural output and its internal distribution, Sakhalin, and Krasnodar krai are from Lisichkin, 1958, pp. 357 and 386. Distribution of the remainder among Komi, Dagestan, and Chechen-Ingush A.S.S.R.'s is based on Balzak, 1949, p. 219.

1940: The distribution by republic is given in postwar handbooks. The distribution within the R.S.F.S.R. is based on data given in Lisichkin, 1958, pp. 304, 312, 333, and 360; Kravchenko, 1963, p. 46; and Trofimuk, 1957, pp. 78, 80–1.

1945, 1950, 1955, 1960: For the postwar period regional distribution of petroleum output by Union republic is available in the statistical handbooks. There is little absolute and solid information on the distribution of petroleum production within the R.S.F.S.R., but many statements concerning the percentage distribution within the R.S.F.S.R. and within other aggregates that include parts of the R.S.F.S.R. The approach used here was to determine the distribution by regions in 1955 and then to project these amounts forward and backward, using indexes for petroleum output. Regional output indexes are scattered, but there are a great many of them, especially in the oblast' statistical handbooks. The distribution for 1955 is based primarily on Brenner, 1957, p. 100, and an article in *VE,* 1957:10. Less detailed percentage distributions for additional years provide some control totals, for 1950 and 1960. The major sources are Brenner, 1962, pp. 51–4; and *Promyshlennost' i stroitel'stvo,* vol. 2, p. 87. There are numerous statements on absolute output and relative importance of the Volga-Ural region: Keller, 1958, p. 15; *Malaia Sov. Entsiklopediia,* 3rd ed.; Trofimuk, 1957, p. 95; Brenner, 1962, p. 54; Umanskii, 1965, p. 105. These provide additional control totals.

This procedure leads to some inconsistencies. It makes output in later years in the Bashkir A.S.S.R. too large to be accommodated with another set of data. From *Nar Khoz SSSR,* a series can be assembled for the years since 1958 for output in the "eastern regions," a term which covers Central Asia and Kazakhstan, Sakhalin, and the Ural region. There are other inconsistencies and puzzles in the evidence as it appears in Soviet sources, but I have tried to generate a complete regional distribution from it that would accommodate the evidence more or less in proportion to its explicitness and my estimate of its reliability. In my original estimation, I did this also for the years intermediate to those in the table, since there is also a great deal of evidence on those years. It should be noted that the distribution within the R.S.F.S.R. has not always been made completely consistent with the control totals. These figures must clearly be taken as approximations, but this is the best that can be done until the Russians publish complete data in absolute amounts for the years 1945–63. For 1964 and 1965, a complete and reasonably consistent regional distribution can be put together from the statistical handbooks and from *EG,* 1966:9, p. 10, supplemented with information on Belorussian and Western Siberian production from *GNIG,* 1966:2, p. 2, and *NKh,* 1966:4, p. 5.

TABLE A-10. SELECTED INDICATORS RELATING TO PRODUCTION METHODS

Item	1940	1945	1950	1955	1958	1959	1960	1961	1962	1963	1964	1965
Production by type of lift (*per cent of total*)												
Flowing wells	23.3	13.7	32.5	58.3	70.4	72.7	73.7	74.0	73.2	70.3	66.8	64.4
Pumped wells	38.4	43.7	44.7	34.0	25.6	24.2	23.2	23.5	24.5	27.4	30.9	33.5
Electric	none	none	neg.	4.0	5.0	n.a.	6.1	7.0	8.4	11.0	13.1	n.a.
Gas lift	37.2	40.9	21.1	6.5	3.4	2.6	2.3	2.1	1.9	1.9	1.9	1.8
Other	1.1	1.7	1.7	1.2	.6	.5	.5	.4	.4	.4	.4	.3
Output from pressure maintenance fields (*per cent of total*)	neg.	neg.	23.0	57.3	n.a.	n.a.	66	65.2	n.a.	70	67	68.3
Water injected (*million m³*)	neg.	n.a.	8.9	81.6	141.6	164.7	189.4	201.5	219.9	263.6	331.5	328.9
Air and gas injected (*million m³*)	neg.	n.a.	295.2	785.8	n.a.	n.a.	563.1	566.4	n.a.	n.a.	479.5	n.a.
Number of hydrafrac treatments	none	none	none	878	2,713	2,262	2,707	2,493	2,140	2,049	2,047	2,083
Output per well-month (*tons*)	201	145	170	213	309	340	371	405	444	469	490	n.a.
Old wells	819	132	153	190	283	315	349	386	421	452	n.a.	n.a.
New wells	173	604	510	866	1,024	1,089	1,047	1,106	1,215	1,064	n.a.	n.a.
Total well-months (*millions*)	155	134	223	332.4	366.4	381.0	398.5	410.0	419.5	439.4	454.5	n.a.
Implied average number of oil wells	14,140	12,462	19,265	28,754	31,680	32,947	34,480	35,475	36,290	38,014	47,970	n.a.
Number of pressure and control wells	neg.	neg.	567	2,425	3,560	n.a.	3,062[a]	3,376[a]	n.a.	n.a.	3,423[a]	3,785[a b]
Number of gas wells[c]	42	n.a.	381	550	717	729[d]	1,147	n.a.	n.a.	1,305[d]	2,079	n.a.
Number of oil wells[c]	20,450	n.a.	n.a.	n.a.	n.a.	38,580	33,943	n.a.	35,694	n.a.	n.a.	46,100
Output divided between												
Old wells (*million tons*)	25.6	n.a.	32.7	60.3	98.3	114.4	135.4	154.0	n.a.	n.a.	n.a.	n.a.
New wells (*million tons*)	5.5	n.a.	5.2	10.4	14.9	15.2	12.8	12.0	n.a.	n.a.	n.a.	n.a.

[a] Pressure wells only. [b] Planned figure. [c] End of year. [d] Annual average.

Sources: Production by type of lift: Nar khoz SSSR, except electric submersibles: *NKh*, 1961:10, p. 6, 1963:11, p. 3, and 1965:6, p. 6; Brenner, 1957, p. 70; Gur'evich, *et. al.*, 1958, p. 174; *VE*, 1955:10, p. 60.

Water, air, and gas injected: Nar khoz SSSR, various years; Lisichkin, 1958, p. 164; Alekhin, 1957, p. 11; *GNIG*, 1966:7, p. 2; Mel'nikov, p. 38.

Number of hydrafrac treatments: Nar khoz SSSR, various years.

Output per well-month: Mel'nikov, p. 42.

Implied annual average number of oil wells: Derived by dividing total output by total number of well-months worked, with a well-month taken arbitrarily as 30 well-days. The first step is to divide output per well-month into output to get well-months worked, multiply by 30 to get well-days worked, and divide by 365 to get well-years. This is time worked, not calendar time. In recent years active wells are said to have worked approximately 95 per cent of calendar time and the figure obtained above is divided by 95 to get actual well years for active wells. In 1940, the coefficient was only 90 per cent, and this has also been used for 1945.

Number of pressure and control wells: Dunaev, I, 1957, p. 60, and II, 1961, p. 46; Mel'nikov, p. 38.

Number of gas wells: 1950: Nikolaevskii, 1961, p. 10; 1955, 1959: Dunaev, II, 1961, p. 46; 1963: Urinson, p. 50; 1964: *GP*, 1966:6, p. 2.

Number of oil wells: Soviet writers are very careless and ambiguous in their statements about the number of oil wells. The principal factors confusing the issue are failure to distinguish between gas wells, oil wells and service wells, and carelessness about the concept used. The broadest concept for Soviet oil wells (and perhaps for gas wells also) is the "exploitation fund" of wells. Within this there is a division into "active" and "inactive" wells. The latter seem to be primarily wells undergoing capital repair. Even within the active category, there seems to be considerable idle time—in recent years "active" wells have generally worked only about 95 per cent of calendar time. Definitions for the various concepts are given in Murav'ev, *Spravochnik po dobyche nefti*, vol. III, pp. 532–3, and in Dunaev, I, 1957, pp. 65–6. There is a great deal more information on the number of wells than that shown in the table, but because of confusion on these points I have not been able to make sense of it.

The figure of 20,450 wells shown for January 1, 1940, corresponds to the "exploitation fund" and is cited in Brenner, 1962, p. 193. The 38,500 shown for January 1, 1960, is not precisely described in the source (Brenner, 1962, p. 195) but must include gas and service wells. The numbers shown for January 1, 1961 and 1963, are the numbers of oil wells producing at that time. (The encyclopedia *Promyshlennost' i stroitel'stvo*, vol. 2, p. 79.) Of the 35,694 oil wells active at the end of 1962, 7,187 were producing by free flow, 27,396 were being pumped, and 1,111 were being produced by gas lift.

The bulk of Soviet oil wells are accounted for by the R.S.F.S.R. and Azerbaidzhan. Of the total 38,500 shown for the end of 1959, 15,952 were in the R.S.F.S.R., 15,005 in Azerbaidzhan. The Turkmen S.S.R. had 1,300, Kazakhstan 2,462, the Ukraine 1,767, and the Uzbek S.S.R. 1,290. A more detailed breakdown is given in Kashnitskii, 1965, p. 62.

Output divided between old and new wells: 1940, 1950, 1955: Keller, 1958, p. 19 and Dunaev, I, 1957, p. 63; 1960–61: *Ekonomika neftedobyvaiushchei promyshlennosti*, p. 200.

Output from pressure maintenance fields: NKh, 1959:9, p. 5; 1961:1, pp. 3–4; 1955: Umanskii, 1965, p. 21; 1963: *NKh*, 1963:11, p. 2; 1964: Umanskii, 1965, p. 52; 1965: *EG*, 1966:9, p. 11; 1950, 1961: Mel'nikov, p. 38.

TABLE A-11. COST OF OIL EXTRACTION BY REGION, SELECTED YEARS

(*rubles per ton*)

Region	1950	1955	1960	1963
U.S.S.R. average	6.24	4.92	3.30	3.04
R.S.F.S.R.	4.58	2.84	n.a.	2.16
Volga-Ural	n.a.	2.06	1.64	1.77
Tatar A.S.S.R.	3.39	1.57	1.37	1.40
Bashkir A.S.S.R.	2.97	1.97	2.10	2.31
Kuibyshev oblast'	2.68	2.21	1.34	1.52
Volgograd oblast'	9.62	1.53	1.46	1.67
Perm oblast'	n.a.	n.a.	4.22	2.37
Orenburg oblast'	n.a.	5.64	n.a.	3.95
Saratov oblast'	n.a.	n.a.	n.a.	3.68
North Caucasus	n.a.	5.35	6.04	4.53
Krasnodar krai	4.70	5.64	4.56	5.75
Chechen-Ingush A.S.S.R.	5.64	n.a.	7.07	4.35
Dagestan A.S.S.R.	9.81	18.50	19.70	8.91
Stavropol'skii krai	n.a.	n.a.	2.13	2.10
Komi A.S.S.R.	n.a.	n.a.	10.20	6.38
Sakhalin	15.24	14.76	11.19	10.18
Azerbaidzhan	6.85	8.86	7.17	7.81
Georgian S.S.R.	53.37	35.92	32.80	28.60
Kazakhstan	11.21	9.55	8.84	10.24
Turkmen S.S.R.	5.30	6.19	4.93	4.44
Fergana Valley	3.50	5.44	6.47	8.38
Ukraine	23.70	12.21	4.32	3.16

Sources: The figures in the table probably actually refer to the cost per ton of oil and oil-well gas together. In their cost accounting the Russians sum the two kinds of output, using heat content as the common denominator, and figure the cost per ton of this composite output.

The starting point for the postwar data is 1955. An absolute figure for cost in the Ukraine is given in Gonta, p. 64, and a number of interregional cost relatives given in Brenner, 1957, pp. 104–5, and in Zasiadko, p. 33, make possible the reconstruction of the whole structure, except for Krasnodar krai, which is based on an index applied to an absolute figure for 1958. The figure for the North Caucasus is figured as a residual. There are two independent checks; for the Tatar A.S.S.R., a figure quite close to that in the table is given in one of the statistical handbooks for the Tatar A.S.S.R., and a figure for the Kazakh S.S.R. close to that in the table is given in Shaukhenbaev, p. 226.

For 1950, indexes of cost trends given in Zasiadko, p. 34, are applied to the 1955 absolute figures. Gonta and Shaukhenbaev also provide some close checks here.

For 1960, several statements about cost trends for the U.S.S.R. imply an all-Union average of about 33 old rubles per ton (*NKh*, 1961:9, p. 13, and Kalamkarov, 1961, p. 8, for instance). Also the cost of oil in the Tatar A.S.S.R. is given as 13.74 old rubles in *Tatarskaia neft'*, 1961:3, p. 1. Most of the rest of the figures for 1960 are worked out using interregional cost relatives given in Brenner, 1962, p. 373 and cost indexes given in *NKh*, 1961:3, p. 14 and L. M. Umanskii, 1962, p. 97, except for Perm, Komi, Sakhalin, Georgian S.S.R., Fergana Valley, and Ukraine, which are from Kantor, p. 75.

Cost for the Fergana Valley was figured as a residual for 1955 and extended to other years using the cost indexes for the Uzbek S.S.R.

The 1963 column is based on cost indexes linking all-Union cost in 1955 and 1963, and a set of regional relatives given in Kashnitskii, p. 132.

Table A-12. Selected Indicators Relating to Transport of Oil and Products

Item	1940	1950	1955	1958	1961	1962	1966
Shipments (*million tons*)	66.4	86.2	166.7	255.0	367.7	413	574.7
Pipelines	7.9	15.3	51.7	94.9	144.0	165.1	247.7
Crude oil	6.7	12.7	45.4	n.a.	128.9	147.3	225.6
Products	1.2	2.6	6.3	n.a.	15.1	17.8	22.1
Railroads	29.5	43.2	77.6	112.5	168.4	190.5	240.2
Crude oil only	n.a.	14.1	15.6	24.8	43.3	51.8	n.a.
River	9.7	11.9	14.4	16.1	20.5	21.2	26.9
Sea	19.6	15.8	23.0	31.5	34.8	36	59.9
Coasting trade only	19.5	n.a.	n.a.	24.4	n.a.	n.a.	n.a.
Turnover (*billion ton-km*)	66.7	80.8	154.5	253	387.8	436	728.4
Pipelines	3.8	4.9	14.7	33.8	60.0	74.5	165.0
Railroads	36.4	52.0	101.6	154.0	230.6	252.5	301.9
Crude oil only	n.a.	12.0	17.0	37.2	73.8	84.9	n.a.
River	12.1	12.0	14.3	15.6	20.6	20.8	30.5
Sea	14.4	11.9	23.9	50	76.7	88.2	231.0
Average length of haul (*km*)							
Pipelines	480	320	284	356	417	451	661.1
Railroads	1,234	1,205	1,309	1,369	1,369	1,325	1,257
Crude oil only	837	851	1,087	1,498	1,705	1,641	n.a.
River	1,259	1,018	1,000	1,013	1,005	981	1,516
Sea	735	752	1,037	1,517	2,204	2,450	3,856
Length of pipelines (*thousand km, Jan. 1*)	4.1	5.4[a]	8.1	13.2	17.3	20.5	28.2
Crude oil	n.a.	n.a.	4.5	n.a.	12.7	15.5	n.a.
Products	n.a.	1.5	3.6	n.a.	4.6	5.0	n.a.

[a] 1951.

Sources: The data in this table are mostly from standard statistical handbooks, but gaps in those sources have been filled in from the following: Galitskii, 1965, p. 97; Gurevich *et al.*, 1958, pp. 283–4; Koldomasov, 1963, p. 312; Sheiman, 1962, pp. 49, 53; Koriakin, 1961; Bakaev, 1961, II, p. 25; E. D. Khanukov, *Transport i razmeshchenie proizvodstva*, 1955, p. 207; *Zh-dt*, 1964:3, p. 63; Markova and Smirnov, 1966, pp. 7–9; and Mel'nikov, 1966, p. 53.

TABLE A-13. ESTIMATION OF SOVIET CONSUMPTION OF PETROLEUM PRODUCTS
(*million tons*)

Year	Refinery runs	Less: Losses	Less: Refinery fuel	Less exports	Add imports	Apparent consumption
1950	34.27	1.53	3.61	0.78	1.2[a]	29.55
1955	64.54	2.58	6.94	5.09	3.82	53.75
1956	77.04	3.08	8.10	6.17	3.80	63.49
1957	88.57	3.55	9.11	7.76	2.94	71.09
1958	99.35	3.97	9.98	9.05	3.22	79.57
1959	111.30	4.45	11.12	12.89	3.34	86.18
1960	123.71	4.95	11.89	15.39	3.23	94.74
1961	134.75	5.39	11.72	17.83	2.74	102.55
1962	152.24	6.09	11.87	19.10	2.32	117.50
1963	167.20	6.69	11.54	21.14	2.34	130.17
1964	175.23	7.01	10.51	19.93	2.08	139.86
1965	186.81	7.47	11.21	20.99	1.90	149.04

[a] Computed by subtracting the imports of crude estimated in Table 22 from the reported figure for products and crude combined.

TABLE A-14. SOVIET CONSUMPTION OF PETROLEUM PRODUCTS, BY PRODUCT
(*million tons*)

Product	1946	1950	1955	1956	1958
Gasoline	3.78	5.55	11.55	14.10	18.86
Auto gas	2.3[a]	4.5[b]	10.34[c]	12.4[d]	n.a.
Kerosene	4.78	6.06	9.46	10.03	12.97
Diesel	0.8	2.30	8.71	10.79	16.07
All light products	9.4	13.9	29.72	34.92	47.90
Residual fuel oil	7.20	10.87	17.47	21.14	25.38
Furnace grade	n.a.	9.0[e]	15.00[f]	17.1[d]	21.34[e]
Naval grade	n.a.	1.90[e]	2.47[e]	4.0[f]	4.04[e]
Other	2.36	4.76	6.56	7.43	6.29
Lube oils	n.a.	1.4[e]	2.1[e]	n.a.	3.3[e]
Asphalt	n.a.	0.8[e]	1.8[e]	n.a.	2.9[e]
Total	18.9	29.55	53.75	63.49	79.57

[a] Index, 1955/1946, Keller, 1958, p. 50.
[b] Index, 1955/1950, Keller, 1958, p. 50.
[c] Deduced from statement by Keller, 1958, p. 50, that in 1955, 38.4 per cent of all consumption of light products was automotive gasoline.
[d] *NKh*, 1957:1, pp. 19–21.
[e] *Khim*, 1959:9, article by Beider.
[f] Figured as residual.
These results are generally consistent with Soviet statements about growth indexes for consumption of individual products, about the ratio of diesel to all light products, etc., although there are many contradictions in detail. It is impossible to reconcile fully all the evidence.

TABLE A-15. SOVIET GAS OUTPUT

(billion cubic meters)

Year	Natural gas output utilized: Gas-well	Oil-well	Total	Manufactured gas	Total	Oil-well gas wasted	Per cent of oil-well gas used
1940	0.4	2.9	3.2	0.2	3.4	n.a.	n.a.
1945	n.a.	n.a.	3.3	0.1	3.4	n.a.	n.a.
1950	3.6	2.2	5.8	0.4	6.2	3.8	56.9
1951	3.8	2.4	6.3	0.6	6.8	n.a.	n.a.
1952	3.9	2.5	6.4	1.0	7.4	n.a.	n.a.
1953	4.4	2.5	6.9	1.0	8.0	n.a.	n.a.
1954	4.9	2.6	7.5	1.3	8.8	n.a.	n.a.
1955	5.9	3.1	9.0	1.4	10.4	2.7	53.6
1956	8.3	3.8	12.1	1.6	13.7	3.3	53.6
1957	14.1	4.5	18.6	1.7	20.2	3.3	57.6
1958	22.5	5.6	28.1	1.8	29.9	4.3	56.0
1959	28.6	6.8	35.4	1.9	37.3	5.4	55.6
1960	37.2	8.1	45.3	1.9	47.2	n.a.	n.a.
1961	50.4	8.6	59.0	1.9	60.9	n.a.	n.a.
1962	63.5	10.0	73.5	1.7	75.2	7.4	57.2
1963	77.7	12.2	89.8	1.7	91.5	19.9	61.2
1964	94.4	14.2	108.6	1.7	110.3	21.4	66.5
1965	111.2	16.5	127.7	1.7	129.4	n.a.	n.a.
(1965)[a]	132.9	15.4	148.3	1.7	150.0	19.6	78.4
(1970)	200–210	n.a.	n.a.	n.a.	225–240	n.a.	n.a.
(1980)	n.a.	n.a.	n.a.	n.a.	700	n.a.	n.a.

[a] Seven Year Plan goal.

TABLE A-16. SOVIET NATURAL GAS RESERVES, BY REGION AND BY CATEGORY, JANUARY 1, 1966

(billion cubic meters)

Region	A + B	C_1	$C_2 + D_1$	D_2
U.S.S.R. total	2,021	1,545	29,364	33,750
R.S.F.S.R.	928	767	23,136	20,930
Arkhangel oblast'	none	none	none	300
Komi A.S.S.R.	9	30	518	800
Perm oblast'	6	18	225	280
Bashkir A.S.S.R.	5	26	146	200
Kuibyshev oblast'	6	5	71	70
Orenburg oblast'	16	9	234	630
Saratov oblast'	41	29	843	660
Volgograd oblast'	68	22	221	1,200
Astrakhan oblast'	2	neg.	100	180
Kalmyk A.S.S.R.	21	26	212	200
Rostov oblast'	none	4	100	200
Krasnodar krai	379	86	933	820
Stavropol'skii krai	192	43	447	100
Chechen-Ingush A.S.S.R.	6	3	50	80
Dagestan A.S.S.R.	10	33	544	130
Western Siberia	149	306	12,293	3,800
Eastern Siberia	8	90	5,803	11,000
Sakhalin	11	38	396	280
Ukraine	448	207	1,034	2,130
Azerbaidzhan	27	27	398	1,100
Kazakh S.S.R.	4	88	1,041	3,400
Turkmen S.S.R.	125	251	1,494	4,620
Uzbek S.S.R.	484	182	1,863	560
Tadzhik S.S.R.	1	12	104	300
Kirgiz S.S.R.	4	10	102	20
Georgian S.S.R.	none	none	100	300
Armenian S.S.R.	none	none	60	60
Moldavian S.S.R.	none	none	30	70
Belorussian S.S.R.	none	none	none	200
Lithuanian S.S.R.	none	none	none	30
Latvian S.S.R.	none	none	none	30

Source: GP, 1967:1, pp. 12–13.

TABLE A-17. GROWTH OF NATURAL GAS RESERVES, CATEGORIES A + B, 1940–66

(billion cubic meters)

		Gross annual increments	
Year	Stock on Jan. 1	Seven Year Plan	Actual
1940	15.2		
1946	36.5		
1947	58.8		
1948	64.9		
1949	71.9		
1950	85.0		
1951	120.7		
1952	223.0		
1953	247.0		
1954	342.0		
1955	389.0		
1956	491.0		
1957	588.0		
1958	700.0		
1959	988.0	359	706.0
1960	1,667.1	391	273.7
1961	1,854.8	410	209.2
1962	2,014.9	425	15.4
1963	1,942.1	445	239.4
1964	2,105.2	480	94.3
1965	2,090.8	530	196.7
1966	2,126.5		

Sources: Stock of reserves: Bokserman, 1964, p. 17, except 1959 and 1964 from *GP*, 1964:5, p. 3; 1965 from *GP*, 1965:8, p. 1; 1966 from *GP*, 1966:4, p. 2, and 1946 from *Problemy nefti i gaza*, 1959, p. 5; Seven Year Plan control figures: Vasil'ev, *et al.*, 1961, p. 13; Actual increments: *GP*, 1966:1, p. 6, and 1966:4, p. 1. The figures shown for increments differ from those available in some other sources, and are not fully consistent with the data on stocks and production, but they seem the best set available.

TABLE A-18. GROWTH OF THE SOVIET GAS PIPELINE NETWORK

Year	Length of pipeline in operation, January 1 (1,000 km) — Transmission lines: Total	Transmission lines: Glavgaz	Gathering lines	Distribution lines	Installed compressor capacity, January 1 (*1,000 hp*)	Gas transported (*billion m³*)
1950	2.3					
1951	2.9					
1952	2.2					
1953	3.3					
1954	4.1					
1955	4.3					
1956	4.9					
1957	7.3					
1958	9.6				102	13.3
1959	12.2	6.6	1.3	10.6		18.2
1960	16.5	12.6	1.7	12.4	109	26.0
1961	21.0	16.5	2.3	12.8[a]		37.4
1962	25.3			17.6[a]		50.7
1963	28.5	24.4			1,270	71.7
1964	33.0	29.9			1,360	87.5
1965	36.9	33.8			2,000	102.9
1966	41.0	38.3			2,378	119.8

[a] The 1962 figure clearly refers to a larger concept than the figures for the other years; and under this concept the 1961 figure is 14.6.

Sources: Gas transported: *GP*, 1964:7, p. 2, and 1966:1, p. 7. Total transmission lines: *GP*, 1966:1, p. 7. Glavgaz transmission lines: *GP*, 1961:10, p. 14, and 1962:12, p. 28; *Stroitel'stvo truboprovodov*, 1964:2; Smirnov, 1962, p. 48. Gathering lines: *GP*, 1961:10, p. 14, and 1962:12, p. 28. Distribution lines: *GP*, 1961:10, p. 14, and 1962:12, p. 28; Furman, 1963, p. 23. Installed compressor capacity: *GP*, 1963:1, p. 2; 1963:11, p. 2; 1965:1, p. 2; 1966:3, p. 21; and Bokserman, 1964, p. 144.

TABLE A-19. COMPARATIVE PIPE SIZE DISTRIBUTION FOR U.S. AND SOVIET GAS PIPELINE NETWORKS, JANUARY 1, 1963

Pipe diameter (*inches*)	Per cent of country total
U.S.S.R.:	
15 and under	25.6
16.7–20.8	26.4
28.3–40.0	48.0
United States:	
10.0 and under	30.5
10.1–15.0	11.6
15.1–20.0	18.3
20.1–25.0	14.3
25.1–30.0	22.7
30.1 and over	2.7

Sources: Federal Power Commission, *Statistics of Natural Gas Companies*, 1962, p. xix; and Bokserman, 1964, p. 70.

TABLE A-20. CONSUMPTION OF GAS BY SECTOR

(*billion* m^3)

Sector	1958	1959	1960	1965	1966	1970 projection[b]
Household-municipal	2.93	4.35	5.69	14.89	15.95	25.00
Industry	14.27	19.46	25.27	74.93	80.62	131.40
Chemicals	0.27	0.96	1.90	6.15	6.32	15.00
Metallurgy	1.70	2.67	5.05	18.43	22.52	43.30
Cement	1.52	2.04	2.91	6.70	8.00	13.00
Machinery and metal-working	1.82	2.79	3.39	12.84	15.21	19.00
Oil and gas[c]	5.64	5.89	5.95	n.a.	n.a.	n.a.
Construction materials and construction	0.86	1.43	1.57	7.26	6.50	5.00
Light industry	0.18	0.59	0.60	n.a.	n.a.	n.a.
Food industry	0.60	1.04	1.21	n.a.	n.a.	n.a.
Other	1.70	2.06	2.70	n.a.	n.a.	n.a.
Electric stations	9.61	10.41	12.23	35.68	40.65	60.00
Transport	0.13	0.12	0.21	0.37	0.82	1.20
Agriculture	0.03	0.04	0.07	0.23	0.46	3.00
Own needs and losses	0.91	0.78	1.59	2.85	4.24	9.60
Export[d]	0.21	0.22	0.24	0.39	0.80	2.30
Total	28.09	35.38	45.30	129.34	143.54	232.50

[a] Manufactured gas excluded before 1965, included in 1965 and after. Manufactured gas amounted to 1.7 billion m^3 in 1965, and approximately the same in other years. In 1958 consumption of manufactured gas was distributed as follows: Household and municipal, 56.9 per cent; industrial uses, 25.9 per cent; electric stations, 11.1 per cent; and miscellaneous, 6.1 per cent (Bakulev, 1961, p. 152). This pattern has changed little in later years.

[b] These figures are clearly a very rough approximation not easily reconcilable with planned 1970 output or with distributions given in other sources.

[c] Includes gas used for production of carbon black (see *GP*, 1961:10, p. 5).

[d] Consists of gas exported from Western Ukraine to Poland.

Note: Data for 1958–60 are from *GP*, 1962:12, p. 4; remainder from *GP*, 1967:2, pp. 2–3.

TABLE A-21. PRODUCTION OF LIQUEFIED GASES, 1958–67

(*thousand tons*)

Year	Total	In petroleum refineries	In gas processing plants
1958	308	195	111
1959	503	290	211
1960	660	336	324
1961	950	494	457
1962	1,256	670	586
1963	1,661	1,070	650
1964	2,299	1,428	950
1965	2,793	1,684	1,204
1967	3,600	n.a.	n.a.

Sources: Total: *NIG*, 1967:1, p. 111, and *GP*, 1967:5. The division into output of petroleum refineries and of gas processing plants is based on an absolute figure for output of petroleum refineries in 1962 from Luzin, 1964, pp. 86–7, and indexes for growth in *NIG*, 1967:1, p. 111. This difference in sources explains why the subseries are not fully consistent with the total.

TABLE A-22. PRICES FOR SELECTED PETROLEUM PRODUCTS IN THE EARLY SIXTIES

(*rubles/ton*)

Product	Zones				
	I	II	III	IV	V
Aviation gasoline					
B100/130	102.50	107.50	110.00	114.00	131.50
B95/130	89.80	94.50	98.60	108.80	123.80
B93/130	87.50	91.50	96.00	106.00	120.00
B91/115	71.50	74.50	78.50	87.50	96.00
B70	62.00	67.00	72.00	79.00	88.00
Automotive gasoline[a]					
A-66	97.00	102.00	108.00	115.00	127.00
A-70 and A-74	112.00	120.00	130.00	142.00	158.00
A-74 and A-76	130.00	140.00	150.00	170.00	185.00
Octane not less than 56	85.00	90.00	95.00	104.00	115.00
Kerosene, illuminating					
Heavy	44.80	44.80	44.80	44.80	44.80
Light	37.00	39.00	41.50	44.00	44.80
Kerosene, tractor					
High octane	32.00	34.00	37.00	40.00	44.00
Regular	29.20	31.70	34.00	36.50	42.00
Naphtha, tractor	37.50	39.50	43.00	48.00	57.00
Fuel oil, marine[b]					
F-12 and F-20	25.00	27.90	29.50	32.00	38.80
Sulfurous, FS-5	24.80	27.70	29.30	31.80	38.60
Fuel oil, grades 20 through 200					
Sulfur content up to 0.5 per cent	24.50	27.40	29.00	31.50	38.30
Sulfur content over 0.5 per cent	18.10	20.40	23.00	25.90	32.00
Auto-tractor diesel fuel					
Summer	29.20	31.70	32.40	35.00	42.00
Summer, sulfur content not more than 0.5 per cent	36.20	38.70	39.40	42.00	49.00
Winter	31.20	33.70	34.40	37.00	44.00
High-speed diesel fuel					
Arctic, winter, and special grades	31.20	33.70	34.40	37.00	44.00
Summer grade	29.20	31.70	32.40	35.00	42.00
Slow-speed diesel fuel	26.20	29.00	30.60	33.10	39.90
Crude oil used as boiler fuel	24.50	27.40	29.00	31.50	38.30
Automotive lube oil					
With special additives	86.00	90.00	95.00	102.00	112.00
Regular	81.00	85.00	90.00	97.00	107.00
Road asphalt					
Grades BN-O and BN-III	21.00	23.00	26.00	31.00	37.00
Grades BN-II-V and BN-III-V	25.00	27.00	30.00	35.00	40.00
Construction asphalt					
Grades BN-IV, BN-V and BN-V-K	27.00	29.00	32.00	37.00	42.50
Road oil, liquid	20.00	22.00	25.00	28.00	33.20

[a] The numbers for different grades of automotive gasoline refer to motor method octane ratings.

[b] Fuel oil for ships (*flotskii mazut*) differs from other fuel oil (*mazut*, *topochnyi mazut*, and *neftianoe toplivo*) primarily by its smaller content of sulfur and other impurities, and by its lower viscosity and freeze point. The various grades of ordinary fuel oil are differentiated by increasing viscosity and rising freeze point. Grade 20 is the least viscous, grade 200 the most.

Source: Voronov, *et al.*, 1962, pp. 300–3. These prices are identical with those shown in the price list published by the Ministry of the Oil Industry, and effective July 1, 1955, except for gasoline. The gasoline prices shown in this table are approximately double those in effect for 1955–58 and after February 1, 1961.

TABLE A-23. DISPOSITION OF SOVIET OIL OUTPUT BETWEEN EXPORT AND DOMESTIC USE

Year	Crude oil output	Exports in field equivalents[a]	Imports in field equivalents[a]	Net exports	Domestic use	Share exported
	(million tons)					(*per cent*)
1950	37.8	1.3	n.a.	n.a.	n.a.	n.a.
1955	70.8	9.3	5.3	4.0	66.8	13
1956	83.8	11.7	6.2	5.4	78.4	14
1957	98.4	15.7	5.0	10.7	87.6	16
1958	113.2	20.6	5.1	15.6	97.6	18
1959	129.6	28.9	5.2	23.7	105.9	22
1960	147.9	37.6	5.2	32.4	115.4	25
1961	166.1	46.4	4.3	42.2	123.9	28
1962	186.2	51.0	3.4	47.7	138.6	27
1963	206.1	57.7	3.4	54.3	151.8	28
1964	223.6	63.0	2.5	57.3	166.3	26
1965	242.9	71.4	2.3	69.1	173.8	28

[a] Products were first converted to a crude equivalent basis on the assumption that refinery fuel expenditure and losses were 15 per cent of the fuel refined; the sum of crude oil and products in crude equivalents was then converted to field equivalents on the basis of an assumed loss rate of 5 per cent.

Table A-24. Geographic Distribution of Soviet Oil Exports

(*thousand tons*)

Product and destination	1955	1956	1957	1958	1959	1960	1961	1962	1963	1964	1965
World total	8,006	10,066	13,681	18,138	25,372	33,218	41,218	45,364	51,382	56,621	64,419
Products	5,090	6,170	7,758	9,045	12,887	15,393	17,830	19,104	21,139	19,930	20,987
Crude	2,916	3,897	5,923	9,093	12,485	17,825	23,388	26,279	30,243	36,691	43,432
Bloc, total[b]	4,198	5,476	7,800	9,278	11,256	15,200	18,450	21,612	22,987	25,295	28,926
Products	1,962	2,553	3,198	3,967	4,891	6,367	8,346	8,938	8,185	7,438	6,486
Crude	2,236	2,923	4,602	5,311	6,365	8,832	10,104	12,674	14,802	17,858	22,440
Asia, incl. China	1,781	1,977	2,104	2,844	3,485	3,373	3,487	3,479	2,129	1,243	752
Crude	378	397	380	689	663	593	24	31	37	40	43
Products	1,403	1,579	1,723	2,156	2,822	2,780	3,463	3,448	2,091	1,204	709
Eastern Europe	2,201	3,157	5,284	2,935	7,324	9,200	10,777	13,283	15,906	18,643	22,397
Crude	1,663	2,216	3,832	4,139	5,289	6,240	7,029	8,716	10,660	13,955	18,292
Products	538	941	1,453	1,796	2,035	2,960	3,748	4,568	5,252	4,688	4,105
Western Europe	2,364	3,341	4,561	5,624	10,297	14,395	16,607	19,035	21,892	22,629	23,833
Products	2,053	2,785	3,589	3,936	5,903	7,380	7,628	9,061	10,266	9,557	9,755
Crude	311	556	973	1,687	8,398	7,015	8,978	9,974	11,626	13,072	14,078
Rest of World (as residual)	1,443	1,250	1,319	3,236	3,820	3,623	6,162	4,737	6,504	8,697	11,660
Products	1,074	831	971	1,143	2,093	1,645	1,856	1,106	2,689	2,936	4,746
Crude	369	419	348	2,094	1,727	1,977	4,306	3,631	3,815	5,762	6,914
Exports to China only	1,589	1,732	1,803	2,507	3,048	2,963	2,928	1,856	1,408	505	38
Crude	378	397	380	672	636	568	0	0	0	0	0
Products	1,211	1,335	1,422	1,835	2,412	2,395	2,928	1,856	1,408	505	38
Residual fuel oil, total	889	1,239	1,697	1,921	3,446	5,343	6,097	6,952	8,576	9,056	9,710
Bloc	16	47	118	148	238	732	1,393	1,515	1,779	2,481	2,524
Eastern Europe	7	39	115	139	221	332	539	1,069	1,351	1,471	1,351
Western Europe	873	1,193	1,578	1,773	3,208	4,611	4,704	5,437	5,894	5,622	5,796
Diesel fuel, total	1,257	1,866	2,464	2,623	3,398	3,897	4,511	5,410	7,488	6,581	7,361
Bloc	503	802	1,058	1,297	1,512	1,874	2,392	2,452	2,377	1,905	1,754
Eastern Europe	245	351	596	545	799	918	1,225	1,577	1,639	1,241	1,243
Western Europe	754	1,064	1,406	1,325	1,886	2,024	2,199	2,958	3,546	3,235	3,083

[a] The figures in this table represent actual exports and so differ from those in Table 43, where actual exports were adjusted to a field-equivalent basis.
[b] Exports to Cuba, Albania, and Yugoslavia are included in the Bloc total only, and not in the regional figures.

Source: Soviet foreign trade handbooks.

TABLE A-25. VALUES AND PRICES IN SOVIET TRADE IN OIL AND PRODUCTS

Item	1955	1956	1957	1958	1959	1960	1961	1962	1963	1964	1965
Average prices realized on exports (*rubles/ton*)											
Crude oil:											
All exports	19.8	19.3	20.2	16.8	15.9	13.9	12.4	12.6	12.9	13.0	12.7
Eastern Europe	29.9	23.7	20.4	19.4	19.8	19.8	19.6	19.5	19.2	18.7	17.1
Western Europe	13.0	14.0	16.6	13.4	12.2	10.3	9.2	8.6	9.1	9.4	9.1
Products											
All exports	29.4	29.3	30.7	25.9	24.2	22.4	22.0	20.7	20.3	18.7	16.6
Eastern Europe	30.7	22.9	30.0	28.2	30.7	31.2	30.4	27.5	27.4	29.1	26.4
Western Europe	21.5	21.3	23.5	18.2	16.3	14.1	13.1	13.0	13.6	12.9	11.8
Prices on products imported from Rumania (*rubles/ton*)	25.0	24.7	27.1	30.7	32.6	32.8	34.0	34.5	34.6	35.1	33.2
Ratio of the price of products to the price of crude oil											
All exports	1.48	1.52	1.52	1.54	1.52	1.61	1.77	1.64	1.57	1.44	1.31
Eastern Europe	1.03	0.97	1.47	1.45	1.55	1.58	1.55	1.41	1.43	1.56	1.54
Western Europe	1.65	1.52	1.42	1.36	1.34	1.37	1.42	1.51	1.49	1.37	1.30
Share of oil in total value of exports (*per cent*)	6.7	7.9	9.1	10.0	10.4	11.8	12.6	11.5	12.5	12.3	12.2

Source: Soviet foreign trade handbooks.

TABLE A-26. OUTPUT OF CRUDE OIL AND NATURAL GAS IN THE SOVIET BLOC

	Crude oil							Natural gas			
Year	Rumania	Hungary	Poland	Bulgaria	Czecho-slovakia	China[b]	Albania	Rumania	Hungary	Poland	Czecho-slovakia
	(- - - -	- - - -	- - - -	thousand tons	- - - -	- - - -	- - - -)	(- - - -	million cubic meters	- - - -	- - - -)
1950	6,211	513	162	none	60	202	132	3,243	379	183	19
1951	8,002	504	181	none	73	305	n.a.	4,045	403	277	n.a.
1952	9,058	600	215	none	113	436	n.a.	4,953	498	313	n.a.
1953	9,741	800	189	none	122	622	n.a.	5,595	547	319	174
1954	10,555	1,200	184	5	125	789	n.a.	5,466	556	358	n.a.
1955	10,920	1,601	180	150	107	966	208	6,169	543	393	173
1956	11,180	1,200	184	247	108	1,163	n.a.	6,756	425	436	274
1957	11,336	675	181	285	108	1,458	490	7,297	411	419	n.a.
1958	11,438	830	175	222	106	2,230	403	8,313	379	384	n.a.
1959	11,500	1,036	175	192	123	3,650	479	9,305	334	424	1,482
1960	11,582	1,215	194	200	137	5,500[c]	728	10,142	340	549	1,439
1961	11,864	1,457	202	207	154	6,708[c]	771	10,914	320	733	1,410
1962	12,233	1,641	202	199	180	7,050[c]	800	13,160	340	821	n.a.
1963	12,395	1,756	212	173	200	n.a.	750	10,101	611	983	n.a.
1964	12,400	1,800	282	160	195	n.a.	764	11,417	784	1,231	n.a.
1965	12,600	1,800	339	229	200	n.a.	825	12,900	1,107	1,378	n.a.

[a] Eastern Germany has negligible output of natural gas or oil, though a large output of manufactured gas (about 3.5 billion cubic meters in the early sixties) and synthetic liquid fuel. Czechoslovakia also has a large output of manufactured gas (nearly 5 billion cubic meters in the early sixties).

[b] The Chinese output includes synthetic, but data for synthetic are available only for certain years, as follows (thousand tons): 1954, 240; 1955, 316; 1956, 407; 1957, 343; 1958, 574; 1959, 597; and 1961, 781.

[c] For recent years the Russians have omitted a figure for Chinese output from their bloc handbooks, implying that they do not know current levels of Chinese output. The figures shown for 1960, 1961, and 1962 are from *NKh*, 1963:5, p. 72, and are also cited in *Khim*, 1964:4, p. 70.

Sources: Statistical handbooks of the countries concerned and Soviet handbooks for the bloc as a whole, except as otherwise noted.

SOURCES

Books

Abramov, M. A., 1971. *Apsheronskii promyshlennyi uzel (The Apsheron Industrial Node),* Baku.

Adamchuk, V. A., *et al.,* 1968. *Problemy razvitiia promyshlennykh uzlov SSSR (Problems of Developing Industrial Nodes in the USSR),* Moscow.

Adamesku, A. A. (ed.), 1973. *Problemy razvitiia i razmeshcheniia proizvoditel'nykh sil Povolzh'ia (Problems of Development and Location of Productive Forces of the Volga Region),* Moscow.

Agaeva, A. A., 1972. *Ekonomicheskaia effektivnost' geologorazvedochnykh rabot na neft' i gaz v Azerbaidzhane (Economic Effectiveness of Geological Exploratory Work for Oil and Gas in Azerbaidzhan),* Baku.

ANSSSR, TsEMI, 1974. *Statisticheskie issledovaniia v otrasliakh narodnogo khoziaistva (Statistical Studies on Branches of the National Economy),* Moscow.

Anufriev, A. F., 1973. *Toplivno-energeticheskaia baza na evropeiskom severovostoke SSSR (The Fuel and Energy Base in Northeast European USSR),* Leningrad.

Avrukh, A. Ia., 1970. *Problemy sebestoimosti i tsenoobrazovaniia v energetike (Problems of Cost and Price Formation in Electric Power),* Moscow.

Baibakov, N. K. (ed.), 1972. *Gosudarstvennyi piatiletnii plan razvitiia narodnogo khoziaistva SSSR na 1971–1975 gg. (The State Five-Year Plan of Development of the USSR National Economy in 1971–1975),* Moscow.

Brenner, M. M., 1968. *Ekonomika neftianoi i gazovoi promyshlennosti SSSR (Economics of the Oil and Gas Industry of the USSR),* Moscow.

Brents, A. D., *et al.,* 1975. *Ekonomika gazodobyvaiushchei promyshlennosti* (Economics of the Gas Extraction Industry), Moscow.

Campbell, Robert W., 1968. *The Economics of Soviet Oil and Gas,* Johns Hopkins University Press for Resources for the Future, Baltimore.

Egorov, V. I., 1970. *Ekonomika neftegazodobyvaiushchei promyshlennosti (Economics of the Oil and Gas Extraction Industry),* Moscow.

Federal Energy Administration. *Project Independence Report: Project Independence,* Washington, D.C., November, 1974.

Feigin, M. V., 1974. *Neftianye resursy, metodika ikh issledovaniia i otsenki (Oil Resources: Methodology of Studying and Evaluating Them),* Moscow.

Feitel'man, N. G., 1969. *Ekonomicheskaia effektivnost' zatrat na podgotovku mineral'no-syr'evoi bazy SSSR (Economic Effectiveness of Expenditures on Preparing the Mineral and Raw Material Base of the USSR),* Moscow.

Gal'perin, V. M., *et al.,* 1968. *Razvitie i perspektivy transporta gaza (Development and Prospects for Gas Transport),* Moscow.

Gankin, M. Kh., *et al.* (eds.), 1972. *Perevozki gruzov (Freight Transport),* Moscow.

Kal'chenko, V. M., 1972. *Gazova promyslovist' i tekhnichnii progress (The Gas Industry and Technical Progress),* Kiev.

Kim, K. M., 1973. *Sovershenstvovanie struktury toplivno-energeticheskogo balansa*

Srednei Azii (Improving the Structure of the Fuel and Energy Balance of Central Asia), Tashkent.

Kortunov, A. K., 1967. *Gazovaia promyshlennost' SSSR (The Gas Industry of the USSR),* Moscow.

Kozyrev, V. M., 1972. *Renta, tsena, khozraschet v neftianoi promyshlennosti (Rent, Price and Economic Accountability in the Oil Industry),* Moscow.

Kulik, G. V., *et al.* (compilers), 1970. *Spravochnik ekonomista kolkhoza i sovkhoza (Collective and State Farm Economist's Handbook),* Moscow.

Lisichkin, S. M. (ed.), 1974a. *Itogi nauki i tekhniki: Ekonomika i organizatsiia proizvodstva otraslei tiazheloi promyshlennosti,* Volume 4, Moscow.

Lisichkin, S. M., 1974b. *Energeticheskie resursy i neftegazovaia promyshlennost' mira (Energy Resources and the Oil and Gas Industry of the World),* Moscow.

Luzin, V. I., 1974. *Ekonomicheskaia effektivnost' i planirovanie kapital'nykh vlozhenii i osnovnykh fondov v neftianoi promyshlennosti (The Economic Effectiveness and Planning of Capital Investments and Fixed Assets in the Oil Industry),* Moscow.

L'vov, M. S., 1969. *Resursy prirodnogo gaza SSSR (Resources of Natural Gas in the USSR),* Moscow.

Melent'ev, L. A., 1973. *Rukovodiashchie ukazaniia k ispol'zovaniiu zamykaiushchikh zatrat na toplivo i elektricheskuiu energiiu (Guidelines for Use of Shadow Prices for Fuel and Electrical Energy),* Moscow.

Mel'nikov, N. V. (ed.), 1968. *Toplivno-energeticheskie resursy (Fuel and Energy Resources),* Moscow.

Mingareev, R. Sh., and V. I. Luzin, 1972. *Ekonomika podgotovki nefti i gaza (Economics of Oil and Gas Preparation),* Moscow.

Oleinik, I. P. (ed.), 1972. *Ekonomicheskoe sotrudnichestvo, sotsialisticheskaia integratsiia i effektivnost' proizvodstva (Economic Cooperation, Socialist Integration and the Effectiveness of Production),* Moscow.

Parashchenko, F. N., 1972. *Regional'nye osobennosti i effektivnost' promyshlennogo proizvodstva na severe (Regional Peculiarities and Effectiveness of Industrial Production in the North),* Moscow.

Pavlenko, A. S., and A. M. Nekrasov (eds.), 1972. *Energetika SSSR v 1971–1975 godakh (Electric Power in the USSR in 1971–1975),* Moscow.

Pobedonostseva, N. N., *et al.,* 1972. *Ekonomicheskaia effektivnost' almaznykh dolot (Economic Effectiveness of Diamond Bits),* Moscow.

Probst, A. E., 1968. *Razvitie toplivnoi bazy raionov SSSR (Developing the Fuel Base of the Regions of the USSR),* Moscow.

Puchkov, N. G. (ed.), 1971. *Tovarnye nefteprodukty, ikh svoistva i primenenie, spravochnik (Commercial Oil Products, Their Characteristics and Application—a Handbook),* Moscow.

Rachevskii, B. S., *et al.,* 1974. *Transport i khranenie uglevodorodnykh szhizhennykh gazov (Transport and Storage of Compressed Hydrocarbon Gases),* Moscow.

Rubinov, N. Z., 1972. *Ekonomika truboprovodnogo transporta nefti i gaza (Economics of Oil and Gas Pipeline Transport),* Moscow.

Savenko, Iu. N., and E. O. Shteingauz, 1971. *Energeticheskii balans (The Energy Balance),* Moscow.

SEV (Council of Mutual Economic Assistance). Secretariat. *Statisticheskii ezhegodnik stran-chlenov Soveta Ekonomicheskoi Vzaimopomoshchi (Statistical Yearbook of the Member Countries of the Council of Mutual Economic Assistance),* Moscow, various years.

Shmatov, V. F., *et al.,* 1974. *Ekonomika, organizatsiia i planirovanie burovykh i neftegazodobyvaiushchikh predpriiatii (Economics, Organization and Planning of Drilling and Oil and Gas Extraction Enterprises),* Moscow.

SSSR v novoi piatiletke: spravochnik (The USSR in the New Five-Year Plan: a Handbook), Moscow, 1966 (no author shown).

Syromiatnikov, E. S., 1970. *Ekonomicheskaia effektivnost' vnedreniia novoi tekhniki v burenii (Economic Effectiveness of Introducing New Technology in Drilling),* Moscow.
Torbin, V. I., 1974. *Territorial'naia differentsiatsiia tsen v tiazheloi promyshlennosti (Territorial Differentiation of Prices in Heavy Industry),* Moscow.
Tovarnye nefteprodukty, svoistva i primenenie, spravochnik (Commercial Oil Products, Their Properties and Uses, a Handbook), Moscow, 1971.
TsSU. *Transport i sviaz' SSSR (Transport and Communication of the USSR),* Moscow, 1972.
TsSU. *SSSR v tsifrakh (The USSR in Figures),* Moscow, 1973.
TsSU. *SSSR i soiuznye respubliki v . . . godu (The USSR and Union Republics in . . .),* Moscow, various years.
Umanskii, L. M., and M. M. Umanskii, 1974. *Ekonomika neftianoi i gazovoi promyshlennosti (Economics of the Oil and Gas Industry),* Moscow.
Urinson, G. S., *et al.,* 1973. *Ekonomika razrabotki gazovykh mestorozhdenii (Economics of Gas Field Production),* Moscow.
U.S. Federal Power Commission. *Statistics of Interstate Natural Gas Pipeline Companies,* Washington, 1972.
Ushakov, S. S., 1972. *Tekhniko-ekonomicheskie problemy transporta topliva (Technical-Economic Problems of Fuel Transport),* Moscow.
VNIGRI (All-Union Geological Exploration Research Institute), 1967. *Trudy,* vypusk 48 (*Works,* part 48). *Geologo-ekonomicheskie issledovaniia poiskovo-razvedochnykh rabot na neft' i gaz (Geological and Economic Studies of Prospecting and Exploratory Work for Oil and Gas),* Leningrad.
VNIGRI (All-Union Geological Exploration Research Institute), 1971. *Trudy,* vypusk 300. *Effektivnost' geologorazvedochnykh rabot na neft' i gaz v SSSR i nauchnye osnovy ikh planirovaniia (Effectiveness of Geological Exploratory Work for Oil and Gas in the USSR and the Scientific Bases for Planning It),* Leningrad.
VNIGRI (All-Union Geological Exploration Research Institute), 1973a. *Trudy,* vypusk 325. Arkhipchenko, A. S. and V. I. Nazarov. *Ekonomicheskaia effektivnost' geologorazvedochnykh rabot na neft' i gaz v zapadno-Sibirskoi nizmennosti (Economic Effectiveness of Geologic Exploratory Work for Oil and Gas in the West Siberian Depression),* Leningrad.
VNIGRI (All-Union Geological Exploration Research Institute), 1974. *Trudy,* vypusk 330. *Effektivnost' poiskovykh i razvedochnykh rabot na neft' i gaz, 1965–1970 (Effectiveness of Prospecting and Exploratory Work for Oil and Gas, 1965–1970),* Leningrad.
VNIGRI (All-Union Geological Exploration Research Institute), 1973b. *Trudy,* vypusk 340. *Ekonomicheskie kategorii geologorazvedochnogo proizvodstva (Economic Categories of Geological Exploratory Production),* Leningrad.
Von Laue, Theodore, 1963. *Sergei Witte and the Industrialization of Russia,* New York.
Zhigalova, I. M., 1973. *Organizatsiia i planirovanie na magistral'nykh gazoprovodakh (Organization and Planning in Gas Transmission Pipelines),* Moscow.

Russian language periodicals

Burenie (Drilling)
Gazovoe delo (The Gas Business)
Neftepererabotka i neftekhimiia (Oil Refining and Petrochemistry)
Neftianaia i gazovaia promyshlennost' (The Oil and Gas Industry)
Organizatsiia i upravlenie neftianoi promyshlennosti (Organization and Administration of the Oil Industry)
Sotsialisticheskaia Industriia (Socialist Industry)
Standarty i kachestvo (Standards and Quality)

Stroitel'stvo truboprovodov (Pipeline Construction)
Transport i khranenie nefti i nefteproduktov (Transport and Storage of Oil and Oil Products)
Trud (Labor)

English Language Periodicals

Current Digest of the Soviet Press
The New York Times
Press Bulletin, Moscow Narodny Bank, London
Soviet News, USSR Embassy, London

INDEX